Energy, entropy, creativity

Reiner Kümmel • Dietmar Lindenberger •
Niko Paech

Energy, entropy, creativity

What drives and slows economic growth

Reiner Kümmel
Institut für Theoretische Physik und
Astrophysik, Campus Süd
Universität Würzburg
Würzburg, Germany

Dietmar Lindenberger
Energiewirtschaftliches Institut (EWI)
Universität zu Köln
Köln, Germany

Niko Paech
Fakultät III, Plurale Ökonomik
Universität Siegen
Siegen, Germany

ISBN 978-3-662-65777-5 ISBN 978-3-662-65778-2 (eBook)
https://doi.org/10.1007/978-3-662-65778-2

Translation from the German language edition: "Energie, Entropie, Kreativität" by Reiner Kümmel et al.,

This book is a translation of the original German edition "Energie, Entropie, Kreativität" by Reiner Kümmel et al., published by Springer-Verlag GmbH, DE in 2018. The translation was done with the help of an artificial intelligence machine translation tool. A subsequent human revision was done primarily in terms of content, so that the book will read stylistically differently from a conventional translation. Springer Nature works continuously to further the development of tools for the production of books and on the related technologies to support the authors.

This Springer imprint is published by the registered company Springer-Verlag GmbH, DE, part of Springer Nature.
The registered company address is: Heidelberger Platz 3, 14197 Berlin, Germany

Striving for greatness means: Control yourself. In self-restraint reveals himself the master. And it's the law alone that gives us freedom.

—Johann Wolfgang von Goethe, 1802 (Vorspiel zur Wiedereröffnung des Theaters in (Bad) Lauchstädt)

Foreword

It is my pleasure to recommend to readers this book, which is driven by the idea of exploring the physical limits of human economic activity using thermodynamics, one of the most universal physical theories we have. But the book does not stop at formulating abstract limits; it transcends limits itself. It gives indications of how consequences can be drawn from inescapable physical laws and how decision-making processes can be prepared in order to shape social developments in harmony with a future-oriented economy according to Western standards. It takes courage to present this to the scientific public.

The economy and society have undergone enormous changes since the late 1970s, when the work pursued by Kümmel, Lindenberger and Paech originated. The overturning influence of the information society—that would perhaps be worth a separate physical consideration—and the onset of globalization were not yet foreseeable in their extent. The rapid development can already be seen in the table of contents, and Reiner Kümmel also takes on issues that, like the energy transition and migration, are no longer in the prediction stage but have long been in the testing stage.

"Is that it?" for the chapter on thermodynamics and economics, or physics and society? Probably not, the challenges for science and society remain, and possibly become greater. I wish the book—where I have been familiar with Reiner Kümmel's ideas for almost 40 years—attentive readers and interest in further questions, also as a personal testimony to his time.

Professor of Physics, University of Bonn
Bonn, Germany
March 2018

Dieter Meschede

Preface

Energy, economic growth, and power, as well as entropy, emissions, and constraints, dominate the current disruption in the world, whose countries are intertwined by mobility and information. The two triples couple thermodynamics with economics.

Thermodynamics describes physical many-particle systems under the impact of energy conversion and entropy production. It was developed to understand the steam engine and its successors steam turbines, gas turbines, gasoline engines, and diesel engines, and to control the processes by which these heat engines, in conjunction with furnaces and reactors, convert the vast energy stores of coal, oil, gas (and uranium) created on Earth by the sun (and cosmic catastrophes) into energy services for humans.

Energy conversion equipment and information processors form the heart of the capital stock of modern industrial economies. Work performance and information processing, which generate the gross domestic product or parts of it, are subject to the First Law of Thermodynamics of the conservation of energy and the Second Law of the increase of entropy, i.e., disorder. However, it is not only hardware and the "constitution of the universe", as which the first two laws of thermodynamics can also be called, that connect thermodynamics and economics but also the property that they are both systems sciences: Thermodynamics studies systems of very many particles interacting with each other via molecular forces, while economics studies the behaviour of economic actors interacting with each other on markets via price-driven trade of goods and services. Systemic similarities lead to the use of similar mathematical methods in thermodynamics and in the theory of economic growth. But this is not the reason why, as a theoretical physicist, I have turned to the problems of energy and economic growth, and energy, emission and cost optimization, in addition to my fields theory of superconductivity and semiconductor theory. Rather, it is the importance of constraints in thermodynamics and economics.

Put casually, the Second Law of Thermodynamics says: Whenever something happens, entropy is produced. If constraints are removed in physical systems, e.g., of their volume or of the energy exchange with the environment, they get out of equilibrium, change, and entropy is produced. When fossil fuels are burned, entropy production includes the emission of heat and particles. This produces environmental pollution of varying severity and intensity unless counteracted by restrictions such as pollution containment and disposal. However, this increases the unavoidable heat emissions and may become problematic in the long term if entropy disposal through heat radiation into space can no longer keep pace with industrial entropy production within the biosphere.

Thus, the Second Law of Thermodynamics imposes constraints on energy-driven economic growth within the biosphere. Even if all particle emissions that are harmful to health and the climate are stopped, the climate-changing heat emissions remain and erect the heat wall.

In the democratically constituted industrialized countries, the actions of economic actors are restricted by frame conditions imposed on the markets by the legislator. Their compliance is monitored by institutions such as antitrust offices, financial, social, health and environmental authorities. This book argues that these frame conditions should be specified and tightened against the background of the fact that energy is a powerful factor of production and that its use is coupled by natural law with the production of entropy that generates emissions.

The book is an update of my book *The Second Law of Economics—Energy, Entropy, and the Origins of Wealth*, also published by Springer. Like that work, it uses parts of my lecture "Thermodynamics and Economics", which I gave between 1990 and 2015 in the Department of Physics and Astronomy at the University of Würzburg, as well as results of the research of my energy working group published in peer-reviewed journals. This entails the limited selection of material, which in no way covers the wealth of literature on energy economics and environmental economics, despite the broadening of horizons through Niko Paech's *post-growth economics*.

The book tries to get by with few equations in the chapters "Entropy and Disruption" (Chap. 1), "Energy and Life" (Chap. 2), "Post-growth Economics" (Chap. 4), "Countries in Disruption" (Chap. 5), and "What Will We Choose?" (Chap. 6). More precise descriptions of facts requiring mathematics are in appendices. The central third chapter, "Wealth Creation and Growth" (Chap. 3), is a quantitative analysis of industrial economic growth in Germany, Japan and the United States. This shows that energy is a powerful factor of production. Here it seems appropriate to present the mathematical methods used to the extent that one can see where, how and why the theory of wealth creation and economic growth described disagrees with the orthodox theory of textbook economics, and that the crux of the matter lies in the consideration of technological constraints. For readers less interested in mathematical reasoning, it tells what can be skipped. Although the chapters refer to each other here and there, they can be read independently.

I am pleased to have Dietmar Lindenberger and Niko Paech as co-authors. Dietmar Lindenberger and I have been working together on energy research for 25 years, and Niko Paech describes in a bluntly realistic way the changes in individual and social behaviour that are likely to be necessary if industrial production is restricted to Earth's biosphere.

As in *The Second Law of Economics*, I want to and must name the people who, not only through their publications but also through intensive personal contacts, have helped me to understand the interaction of thermodynamics and economics and to think about solutions to the social problems that become apparent from the synopsis of these two disciplines. These are the social ethicist Wilhelm Dreier (†), the economists Wolfgang Eichhorn, Alfred Gossner and Wolfgang Strassl, the physico-chemist and energy scientist Willem van Gool (†), the physicist and visionary of space industrialization Gerard K. O'Neill (†), the mathematician Jürgen Grahl and the pioneers of energy and environmental research Charles A. Hall and Robert U. Ayres. I have also received many suggestions from members of the Study Group on Development Problems of Industrial Society (STEIG e. V.) as well as colleagues in the Working Group on Energy of the German Physical Society (DPG) and in the Association of Socio-Economic Systems of the DPG.

I am grateful for the fact that, despite publicistic efforts to the contrary, the unity of research and teaching at German universities has so far been preserved. In the major course lectures and the associated exercises and practical courses in physics, the substantive and methodological foundations of the subject are taught. Elective courses and special courses lead to the front lines of research, often in the field of the teacher. This helps students to choose the area in which they will do their examination papers for the various degrees. These involve independent research under the guidance of a lecturer and often lead to publications in international scientific journals. The results of these are fed back into the appropriate courses. Thus, the unity of research and teaching inspires scientific progress.

The beauty of this is that young people with an alert mind are very interested in new things. They also do not shy away from risks, as interdisciplinary research in particular entails. In the field of thermodynamics and economics, I owe a lot to the following collaborators in the order they joined my energy working group: Klaus Walter, Bruno Handwerker, Helmuth-M. Groscurth, Uwe Schüssler, Thomas Bruckner, Dietmar Lindenberger, Volker Napp, Alexander Kunkel, Hubert Schwab, Julian Henn, Jörg Schmid, Robert Stresing and Tobias Winkler.

I thank my wife Rita for the fact that in the summer of 1968, not long after we had moved to the vicinity of Frankfurt, she looked around our bourgeois, comfortably furnished living room, asked, "And that's it?", recalled the plans that had been nurtured before we had moved to Champaign-Urbana, and thus kicked off the venture that took us to Cali in Colombia. It was there that I began to understand that thermodynamics is important to economics, the environment and society.

Würzburg, Germany Reiner Kümmel
March 2018

Contents

About the Authors

Reiner Kümmel born in 1939, is a retired professor of theoretical physics. His fields of research are the theory of superconductivity, semiconductor theory and energy science. Before his appointment to the University of Würzburg in 1974, he studied/researched/taught at the TH Darmstadt (physics diploma), the University of Illinois at Champaign-Urbana (research assistant to John Bardeen), the University of Frankfurt/M. (doctorate and habilitation) and the Universidad del Valle in Cali, Colombia. From 1996 to 1998 he was chairman of the Working Group on Energy of the German Physical Society.

Dietmar Lindenberger studied economics and physics in Stuttgart, Würzburg and Albany (USA), received his doctorate in 1999 on issues of energy and economic growth and habilitated in 2005 at the University of Cologne. He is active in teaching and research and has consulted, among others the EU Commission, the Federal Chancellery, federal and state ministries, national and international energy companies and research funding institutions. He has published widely on energy issues and is lead author of the energy scenarios for the German government's energy concept.

Niko Paech studied economics, received his doctorate in 1993, habilitated in 2005 and held the Chair of Production and Environment at Carl von Ossietzky University Oldenburg from 2008 to 2016. He currently researches and teaches at the University of Siegen in the Master's programme in Plural Economics. His main research interests are post-growth economics, climate change mitigation, sustainable consumption, sustainable supply chain management, sustainability communication and innovation management. He is active in various sustainability-oriented research projects, networks and initiatives as well as in the supervisory board of two cooperatives.

Entropy and Disruption 1

> The law that entropy always increases holds, I think, the supreme position among the laws of nature. ...if your theory is found to be against the second law of thermodynamics, I can give you no hope: there is nothing for it but to collapse in deepest humilation. (Sir Arthur Eddington, 1929. [2, p. 113])

With the fall of the Berlin Wall and the Iron Curtain between November 1989 and October 1990, some hoped for "the end of history" [3] and the beginning of an age of freedom for people and markets. In its peace the history of conflicts and struggles would find its end. In the meantime, we know how deceptive this hope was. Global welfare requires new constraints. Perhaps the best way to approach this insight is by the seemingly most arduous route—the entropic one. The very word "entropy" frightens many. But entropy can be used to better understand the developmental problems of industrial society discussed in the book than without it. Let us try it.

1.1 Constitution of the Universe

1.1.1 Disorderly Equilibrium

"Entropy – I never get it," is written in the thought bubble above the head of a physicist staring into a textbook on his desk: The desk is covered with a chaotic mess of notes, pens, screwdrivers, cables, files, a matchbox, a coffee cup, a magazine, a calculator, an ashtray and small parts. The cartoon is explained by the text: "Many natural scientists become never really familiar with entropy" [4]. Yet our physicist has it right in front of his eyes. Because

R. Kümmel et al., *Energy, entropy, creativity*,
https://doi.org/10.1007/978-3-662-65778-2_1

Entropy is the physical measure of disorder.

If the pondering thinker now picks up a pen here, a piece of paper there, and puts them down again in other places, the overall state of things has changed somewhat, but the general disorder has not changed at all. The larger the table and the more things that can be distributed on it, the greater the disorder appears. For there are correspondingly more states of distribution, all of which give the impression that their individual parts have just been poured out of a bucket onto the table. Of course, the physicist did not do such a thing. But in the effort to bring order to his understanding of entropy in his brain, the initially well-ordered objects on the table got mixed up. Eventually, things have to be tidied up again. That requires work.

Parents make corresponding experiences with their children. As long as they are little and arrange their ideas of the world and its things in their heads when playing with their toys, however simple these may be, they distribute these toys in the room. The more toys there are and the larger the children's room is, the greater the mess in the evening. The parents' tidying up then ensures that the children find the toys back in their usual places on the shelves the next morning. Even teenagers still seem to enjoy the clutter of notebooks, pens, calculators, friendship bracelets, hair clips, CDs, keys, posters and magazines that they produce while studying for school and relaxing during breaks, sending resigned parents a postcard from abroad with the same mess as in their room, emblazoned with the text: "It's my mess, and I love it."

More physically, the connection between disorder and entropy can be explained using a system of non-interacting particles in a box. Figure 1.1 shows *one* state of such a many-particle system. The state of each *individual particle* is characterized by the length and direction of the momentum arrow attached to it and by the position of the small cell it occupies in the total volume of the box. (For the sake of clarity, only particles in the frontmost cells of the box are shown.) If even one particle changes to

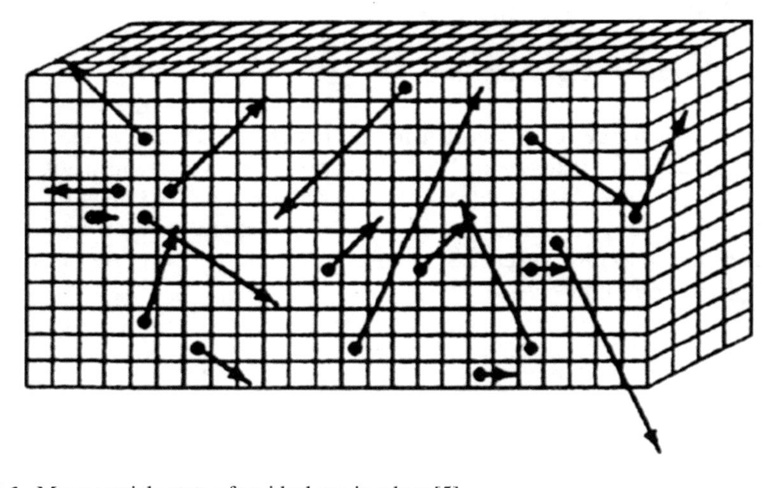

Fig. 1.1 Many-particle state of an ideal gas in a box [5]

another cell, we already have another many-particle state. The number of resulting many-particle states in the box, which look as disordered as in the illustration, is now extraordinarily much larger than the number of possible states in which, for example, all particles are ordered in the cells of a small region of space. (Each of these relatively rare many-particle states corresponds to a possible arrangement of toys on children's room shelves.)

For the sake of simplicity, a state of many particles was chosen for Fig. 1.1, whose interactions with other particles are negligibly small. However, the concepts and laws of thermodynamics, which will be discussed in the following, are also, and especially, valid for systems in which interactions between particles are important.

All attempts by natural scientists to describe the distribution of elementary particles, atoms and molecules and the resulting physical, chemical and technical processes are based on the assumption that a system in *equilibrium* occupies each of the many-particle states accessible to it with equal probability. Since there are overwhelmingly more "disordered" states as in Fig. 1.1 than ordered states within the number Ω of *total* states accessible to the system, disorder in an equilibrium system must increase with this number Ω. Ludwig Boltzmann (1844–1906) showed that this increase is logarithmic, and that the state function entropy S of thermodynamics is statistically given by the equation (written on his tombstone)

$$S = k_B \ln \Omega. \tag{1.1}$$

Here $k_B = 1.3807\ 10^{-23}$ Ws/K is the natural constant named after Boltzmann, and ln is the natural logarithm.

For the following it is of importance that the equilibrium state of a system, and thus its entropy, is determined by the *constraints to* which the system is subject. In Fig. 1.1 these constraints are not shown. One has to imagine them. They are, first, the walls that confine the box. These prevent the particles from moving out of the volume shown. The other is the thermal insulations of these walls, which prevent the particles in the box from exchanging energy with the environment of the box in the form of heat, so that their momentum arrows change. Heat is a *special* kind of energy, namely one that, like work, crosses system boundaries. In general:

Energy is the capacity to cause changes in the world. It is stored in matter and force fields.

In thermodynamic equilibrium, nothing happens—except random fluctuations in particle distribution. An equilibrium system knows no history.[1] Only when the constraints determining the equilibrium are removed, events occur, "history" begins,

[1] So it is also speculated that a closed universe would suffer the "heat death" in a distant future when it should have reached its state of equilibrium. Then matter would be distributed as evenly as possible among all its states of location and motion, and all radiation energy would have the same temperature everywhere. Nothing could happen any more. But who knows whether the universe is closed or open?

for better or for worse. To the good belongs liberation. This is what the next chapter is about. To the bad belong impairments of the natural foundations of life and of interpersonal relationships. This is what the rest of this chapter is about.

1.1.2 Entropy Production and Environmental Pollution

Suppose one of the constraints that keep a system in equilibrium were removed. For example, in our system of Fig. 1.1, the box volume would suddenly be increased by removing the walls that prevented the particles from expanding into a larger volume. Or one removes the thermal insulation of the walls from the outside world, so that the many-particle system absorbs or gives off heat. In either case, the system goes out of equilibrium. And then entropy is produced.

The *empirical fact* that non-equilibrium systems produce entropy is the content of the Second Law of Thermodynamics (formulated mathematically in Eq. (1.3)).

The conservation of energy, the First Law of Thermodynamics and the Second Law of Thermodynamics form the constitution of the universe.

If one wants to restore the original, more orderly state of the system, one must reintroduce the constraints. In our example, this can be done by sucking the particles back into the box, whose walls are back in their original place, by means of an electrically operated pump, or by extracting/supplying the previously absorbed/emitted heat by means of a cooling unit/heating rod and then reapplying the thermal insulation. In all cases, energy must be expended to restore the original order. Experience shows that this further increases entropy in the *overall system* "box + environment including energy suppliers".

Processes that occur after the removal of constraints and produce entropy are called "irreversible", which means that after their devolution the world is no longer the same as before. Only changes that proceeded infinitely slowly could avoid entropy production. But such processes do not exist in real life. That's why (unfortunately) the Second Law of Thermodynamics applies at all times and everywhere, which can also be casually pronounced like this:

Whenever something happens, entropy is produced.

It affects all measurable experiences of change in the world. Any attempt to escape it fails, and any theory that contradicts it is meaningless.

Entropy production in the removal of constraints accompanies the evolution of the cosmos and of life on earth. Progress, especially in economic matters, aims at the removal of constraints. This expands energy use and entropy production. Let us look at how this affects the environment and resources.

Let us consider an industrial country, e.g. the Federal Republic of Germany. It is an open thermodynamic system in which the non-equilibrium processes of life and mechanized value creation take place. It receives energy from the sun and radiates heat into space. Its power plants, factories, automobiles, and heating systems burn coal, oil, and gas, and until their final shutdown in 2022, some nuclear power plants also continued to convert matter into thermal and electrical energy.[2] Goods and services are exported and imported across national borders. The bottom of the national territory limits the system.

Despite the non-equilibrium state, it is possible to investigate the entropy production of technical and economic processes with thermodynamic variables such as pressure and temperature, which are, strictly speaking, only defined in equilibrium systems. For this purpose, it is sufficient that *local* thermodynamic equilibrium prevails in the many small space cells into which the system is (mentally) divided and which are only tiny by the standards of our everyday world but nevertheless contain many billions of atoms and molecules. This is true if, during time intervals in which changes detectable to us occur, an enormous number of "collisions" take place between the atoms and molecules of each cell. This is usually the case.

For every moment one can write down the entropy balance of Germany (or any other part of the earth). For this purpose we imagine that the national borders extend into the three-dimensional up to the height for which rights must be obtained to fly over Germany. At the top edge we hang an immaterial veil that reaches down to the ground, and put an immaterial lid on top. The entropy balance of this thus defined volume of Germany, V, is then: The change per unit time of the total entropy in the volume of Germany, $\frac{\mathrm{d}S}{\mathrm{d}t}$, is equal to the entropy exchanged per unit time t with the outside world, $\frac{\mathrm{d}_a S}{\mathrm{d}t}$, plus the entropy $\frac{\mathrm{d}_i S}{\mathrm{d}t}$ produced per unit time inside the volume:

$$\frac{\mathrm{d}S}{\mathrm{d}t} = \frac{\mathrm{d}_a S}{\mathrm{d}t} + \frac{\mathrm{d}_i S}{\mathrm{d}t}. \tag{1.2}$$

This entropy balance equation corresponds exactly to the balance equation of a savings bank account: The change in the amount of money in the account—it corresponds to $\frac{\mathrm{d}S}{\mathrm{d}t}$—is given by all deposits and withdrawals—corresponding to $\frac{\mathrm{d}_a S}{\mathrm{d}t}$—plus the interest produced in the account, which corresponds to $\frac{\mathrm{d}_i S}{\mathrm{d}t}$.

But an important difference exists between money balance and entropy balance. The interest produced in the account can be positive, namely deposit interest, or negative, namely overdraft interest. But the entropy produced per unit time by non-equilibrium processes in a system is always positive:

$$\frac{\mathrm{d}_i S}{\mathrm{d}t} > 0. \tag{1.3}$$

This mathematical formulation of Second Law of Thermodynamics is based exclusively on *experience*, which was already stated above in words.[3]

[2] Notwithstanding this, and notwithstanding the existence of electric stoves and heaters, a vice-president of the German Bundestag declared in spring 2015 on the occasion of a lecture in Würzburg: "You can't warm rooms with nuclear energy, ladies and gentlemen!".

[3] The Second Law cannot be mathematically disproved, even if a physicist with a doctorate from a prestigious research institute claimed at an international energy conference in Usedom in the early 1990s that he had succeeded in doing so and later repeated this on television in an EU country.

The importance of the Second Law for the environment becomes apparent when Eq. (1.3) is broken down using the balance equations for mass, momentum and energy that are valid in non-equilibrium systems. This is done in Appendix A.1. The result, Eq. (A.4), states:

Entropy production is associated with emissions of heat and particles.

These emissions carry out the commandment of the Second Law to distribute energy and matter as uniformly as possible in the system. In the industrialized planet Earth, this is done by the flows of heat and matter that escape from the furnaces, heat engines, and reactors of the energy conversion plants. These emissions change the energy flows and the chemical composition of the biosphere, to which the living beings and their populations had adapted (more or less) optimally in the course of evolution. If these changes are so severe that they cannot be reversed by the physical, chemical, and biological processes driven by the sun and heat radiation into space, and if they occur so rapidly as to cause health and social adaptation deficits in plant, animal, and human individuals and their societies, the emissions are perceived as pollution.

Air Pollution and Emission Sources

Table 1.1 shows the main air pollutants and the energy sources that emit them. They are sulphur dioxide (SO_2), nitrogen oxides (NO_x), particulate matter, carbon monoxide (CO), hydrocarbons (C_mH_n) and ionising radiation/radioactive particles (radioactive). Geothermal energy, which is not listed, is particularly associated with H_2S emissions if it comes from sources heated by magma.

The use of *all the* energy sources listed also leads to emissions of greenhouse gases, including CO_2 and ozone (O_3) from the combustion of coal, oil and natural gas and their products, and methane (CH_4) from coal mining and the production and distribution of natural gas. Greenhouse gas emissions from uranium, as well as solar, wind, and hydro, are based on the life-cycle CO_2 emissions shown in Table 1.2, which are generated during the production of nuclear power plants, solar cells, solar panels, reinforcements, wind turbines, and dams using fossil fuels; the decay of vegetation in forest-flooded reservoirs of hydropower plants also releases CO_2.

Table 1.1 Emissions from energy sources, excluding greenhouse gases

Energy source	SO_2	NO_x	Dust	CO	C_mH_n	Radioactive
Coal	X	X	X	X	X	X
Oil	X	X	X	X	X	
Natural gas		X		X		
Uranium						X
Sun, wind						
Water						
Biomass	X	X	X	X	X	

Table 1.2 Total, specific life-cycle CO_2 emissions of electrical energy for different energy systems, in grams per kWh. Included are the emissions from construction, operation and maintenance of the systems [7–9]

Lignite-fired power plant	850–1200
Hard coal-fired power plant	700–1000
Gas power plant	400–550
Solar cells	70–150
Hydroelectric power plant	10–40
Nuclear power plant	10–30
Wind farm	10–20

Table 1.3 Emissions from energy conversion and their damaging effects (excluding CO, described in the text, C_mH_n, radioactive substances and greenhouse gases)

Substance	SO_2	NO_x	Dust
Origin	S-content	N-content	Fly ash, traffic
	of coal and oil		
		High combustion temperature	
Conversion	H_2SO_3	HNO_2	
	H_2SO_4	HNO_3	
Effect on humans	Heart, circulation	reduced diffusion of CO	
	Diseases of the respiratory tract		
Ecosystems	Forest damage		
Building	Corrosion		

The production of biomass results in the emission of nitrogen oxides (N_2O) if nitrogenous artificial fertilizer is used, and CO_2 emissions if supported by oil-dependent agricultural technology; without recultivation measures, the emissions from biomass combustion are added to this. Germany now has a very large number of biogas plants in operation—biomass as a whole contributed about two thirds to the 12% contribution of all renewables to meeting Germany's primary energy demand in 2016. Accidents and leakages of methane into the atmosphere are becoming more frequent. Global warming potential of a methane molecule is 25 times that of a CO_2 molecule [6, p. 348].

The estimation of the life cycle CO_2 emissions of solar cells in Table 1.2 takes into account that in 2014 solar cells from Chinese production had captured about two thirds of the world market and that the CO_2 emissions from Chinese solar cell production are about twice as high as those from German production [9].

Damage Effects of Emissions

The direct damage potentials of the emissions listed in Table 1.1 have been known for a long time. Table 1.3 indicates them. Sulphur and nitrogen oxides are irritant gases that increase respiratory resistance. SO_2 causes bronchoconstriction, and NO_x reduces the partial arterial pressure of O_2. In animal experiments with nitrogen oxides (NO_x), destructive processes were found on the surface cells of the alveoli and the bronchial system, as well as an influence on the resistance of the lungs to pathogenic germs. Carbon monoxide (CO), which is not listed in Table 1.3, is formed during incomplete combustion and impedes oxygen transport in the blood. Respirable particulate matter with a particle size of up to 10 μm, which accounts for about 85% of all dust emissions, is likely to be a major factor in bronchopulmonary diseases even at low concentrations. In the London smog disasters of 1948 and 1952, a total of over 4000 more people than normally expected died from pneumonia, heart failure and coronary heart disease. This triggered decisive action against air pollution, which has stopped such smog disasters, at least in the highly industrialized countries. But the construction and operation of the dust filters, as well as the denitrification and desulphurisation plants that convert the flows of dangerous particles and molecules into (nowadays still) harmless heat flows, require energy and capital. Emerging and developing countries that cannot or do not want to raise the funds for this are suffering more and more from the consequences of air pollution as industrialization grows.

In the 1970s and 1980s, the German public was alarmed by a dramatic increase in forest damage caused to a considerable extent by acid rain resulting from SO_2 and NO_x emissions from large German combustion plants ("forest dieback"). By the mid-1980s, environmental awareness and prosperity in Germany had advanced to the point where it was politically possible to introduce statutory emission limits for SO_2 and NO_x through the Large Combustion Plant Regulation. They apply to power plants with a thermal output of more than 300 MW_{th}. According to these limits, a coal-fired power plant with an electrical output of 750 MW_{el}, for example, may not emit more than 6200 t SO_2 and 4100 t NO_x (NO_x, calculated as NO_2) per year. In fact, the power plants did much better. Their SO_2 and NO_x (NO_2) emissions decreased in (West) Germany from 2 and 1 million tonnes in 1980 to 0.3 and 0.5 million tonnes in 1989.

This shows what engineering and industry can do when they have to. It also provides a good example of the beneficent effect of new constraints under the pressure of the Second Law. However, despite all our commitments to environmental and climate protection, we Germans have not yet been able to bring ourselves to introduce a universal speed limit on motorways like all the other civilised countries in the world, say 130 km/h, which would significantly reduce emissions of CO_2, NO_x and particulate matter in the transport sector.

Anthropogenic Greenhouse Effect

A pressing problem of global environmental change is the amplification of the natural greenhouse effect through emissions of greenhouse gases. This leads to an increase in the earth's surface temperature and climate changes. For readers who are still interested in some details on the greenhouse effect and its amplification by emissions from human activities, the effect of greenhouse gases in the atmosphere is described in somewhat more detail in Appendix A.1.2. A more detailed account can be found, for example, in [6] or in the chapters on energy and entropy of [2].

Pope Francis' encyclical *Laudato Sí* [10] from 2015 exhorts humanity in moving words to treat the goods of our common home, the biosphere of our earth, with care.

Even among atheists it met with a positive response, as could be seen in discussions on the fringes of an international energy conference. On the one hand it is good when religion and science approach each other, on the other hand the encyclical contains in its points 23 and 106 passages that show what difficulties the understanding of physical and economic facts causes to the representatives of disciplines and institutions that want to tell people what is good and right.

In point 23 of the encyclical it says about the anthropogenic greenhouse effect (bold emphasis by R.K.):

> The climate is a common good of all and for all ... numerous scientific studies show that most of the global warming of recent decades is due to the high concentration of greenhouse gases (carbon dioxide, methane, nitrogen oxides and others) emitted mainly because of human activity. ***When they intensify in the atmosphere, they prevent the heat of the sun's rays reflected from the Earth from being lost in space***

Comment: Thus, we would all eventually evaporate in the accumulated heat. Correct would be, "As they intensify in the atmosphere, the heat received by the sun's rays can only be radiated back into space as the earth's surface temperature rises."

Now one might still dismiss this comment as the Beckmesserei of a physicist. But the negative judgement of the encyclical under point 106 about technology and industrial society (bold emphasis by R.K.) goes to the substance of the modern production of material prosperity:

> 106. THE GLOBALIZATION OF THE TECHNOCRATIC PARADIGM. The basic problem is ... how humanity has actually embraced technology and its development along with a homogeneous and one-dimensional paradigm. According to this paradigm, a conception of the subject emerges which, in the course of the logical-rational process, gradually embraces the external object and thus possesses it. This subject unfolds by setting up the scientific method with its experiments, which is already explicitly a technique of possessing, mastering, and transforming. It is as if the subject found itself face to face with the formless, which is entirely at its disposal for manipulation. **It has always happened that man has interfered with nature. But for a long time the characteristic was to accompany, to submit to the possibilities offered by things themselves. It was a matter of receiving what the reality of nature offered of itself, reaching out, as it were. Now, on the contrary, the interest is directed towards extracting everything that is at all possible from things through the intervention of man, who tends to ignore or forget the reality of what he has before him. That is why man and things have ceased to join hands in friendship and have become hostile to each other.**

Comment: Romanticizing, unrealistic view of agrarian society and misjudging the opportunities and risks of industrial society. The Old Testament, e.g. 3 Kings 5:27–32; 12:1–17, shows how in an agrarian society "man and things ... joined hands in friendship". Closer examples are the building of St Peter's Basilica, the seizure of the Americas by the European conquerors, and the destruction of forests by clearing for arable land and by felling for energy and building materials. The fate of a country when its economic development gets stuck in the transition from feudal agrarian society to industrial society is illustrated by the example of Colombia in Sect. 5.2.

Rejection of the modern world shaped by natural science and technology can also be found elsewhere among contemporary institutions and individuals who wish to address current social problems. For example, a "Foundation for Cultural Renewal" that emerged from the "Denkwerk Zukunft" announces the start of its operation with the declaration of its board of trustees chairman Meinhard Miegel saying: "Science, art and religion form a harmonious triad in cultures that are fit for the future. In the culture of the West, this harmony is disturbed. The Cultural Renewal Foundation wants to help restore it." Which "sustainable cultures" the West should take as a model is not said. Then conductor Kent Nagano is quoted as saying, "We lose an incredible amount [by focusing on the natural sciences]: inspiration, comfort, public spirit, part of our great Western tradition. We lose the possibility of discovering and experiencing things greater than ourselves ...". [11].

Contemporary thinkers, who are also active in this foundation, address the anthropogenic greenhouse effect as an important problem of the modern world. In this context, the two "assignments of blame" to natural science and technology cited here are by no means exceptional cases. Therefore it may and must be recalled that at the First European Ecumenical Assembly "Peace in Justice" of the Christian Churches of Europe in Basel from 15 to 21 May 1989 members of the Working Group on Energy of the German Physical Society and other natural scientists close to the Christian faith had urgently warned against the anthropogenic greenhouse effect, the Antarctic ozone hole enlarged by CFC emissions and the nuclear winter threatening after a nuclear war. Before and during the Ecumenical Assembly attempts were made in discussions with bishops and cardinals to induce them to take note of the findings of natural science and to take them into account in theological and social-ethical doctrinal statements on creation. Since the mid-1980s, natural scientists have been pointing out that the increasing carbon dioxide emissions from industry, agriculture and the clearing of virgin forests in the course of economic growth are leading to a rise in the earth's surface temperature [12, 13]. Without the findings and warnings from natural science, mankind would be blind and helpless in the face of climate change.

"But climate change is a consequence of the increasing burning of fossil fuels in the course of industrialization, and industrialization requires the natural sciences," you might object. That is true, but the invention of the steam engine, which triggered the Industrial Revolution, took place several decades before physicists and chemists knew what energy and entropy were and before they knew the laws of thermodynamics. It was the answer to Europe's energy crisis in the eighteenth century, when much of the forests had been cleared, firewood was scarce, English coal seams were advancing into deep layers of water, and Newcomen's inefficient steam pumps were no longer sufficient to pump them empty. That is when, in 1765, 29-year-old Scottish instrument maker James Watt had the idea of adding a condensing chamber to Newcomen's steam pump, which greatly increased the machine's energy efficiency and economy. Further improvements and the combination with a flywheel and connecting rod led to the steam engine that powered mechanized looms, locomotives, and steamships. The coal and iron industries expanded. Innovations

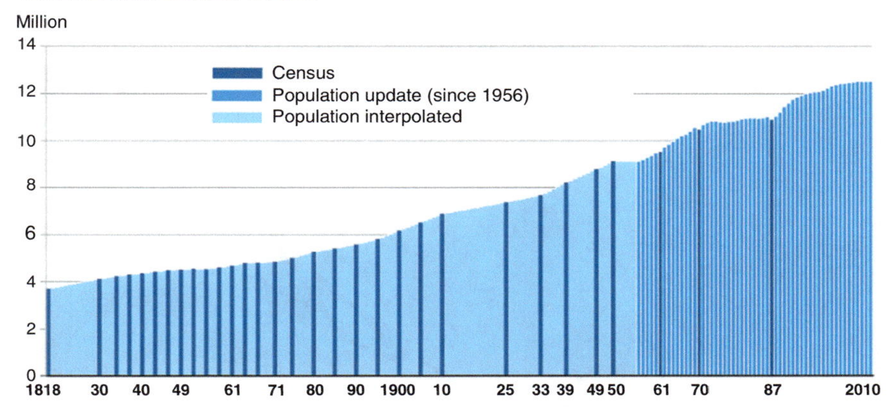

Fig. 1.2 Population growth in Bavaria, the German region whose area has remained virtually unchanged for 200 years [14]

spurred the construction industry. England's industrial boom spilled over into continental Europe and North America [2, S. 16, 17, 48–50].

The (first) Industrial Revolution was the result of a challenge of scarcity and the response of human creativity through invention. The natural sciences were not the primary causal agents. But they are indispensable for understanding and shaping further industrial development, for risk assessments and for new ideas to minimize risk. That is why the humanities and arts should not blame them, but cooperate with them.

According to Fig. 1.2, Bavarian population density more than tripled between 1818 and 2010. Without industrialization, such a densification would not have been possible, while at the same time the standard of living increased enormously. Which aristocrat—or even peasant just released from serfdom—in the early nineteenth century could have even imagined the consumption possibilities of food and culture, e.g. of music on recordings, the living comfort, the medical care, as well as the travel and educational possibilities of the average German citizen of the twenty-first century? And likewise there is no alternative to the industrial society based on science and technology for a humanity that is expected to grow to 10 billion individuals. This is the subject of the following chapters. The only question is under which framework conditions of the market we will do business. Here the right decisions must be made quickly. As Fig. 1.3 shows, annual global CO_2 emissions from the combustion of fossil fuels *alone* rose from 21 to 31 billion tonnes between 1990 and 2010.

The concentration of *all* greenhouse gases—their climate impact being converted to CO_2 equivalents—was already 430 ppm in 2000 [16], see also [17]; meanwhile,

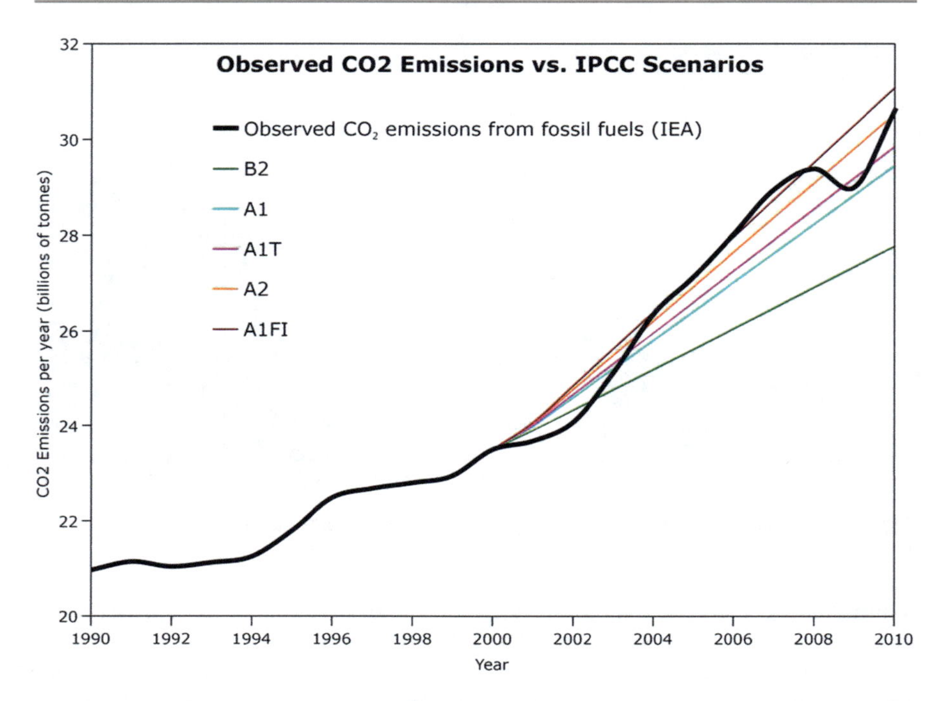

Fig. 1.3 Observed annual global CO_2 emissions, in billions of tons, from fossil fuel combustion between the years 1990 and 2010, thick line, according to the International Energy Agency (IEA), and emission scenarios of the United Nations Intergovernmental Panel on Climate Change (IPCC), thin lines [15]

the news reports a concentration of 400 ppm for CO_2 alone.[4] A doubling of the atmospheric CO_2 concentration compared to its pre-industrial value of 280 ppm and an associated, perhaps just tolerable, increase in the earth's surface temperature of about 2 °C is likely to be achieved within the next few decades. In order for this temperature increase not to be exceeded, global annual CO_2 emissions would have to be approximately halved by 2050 compared to 1990, and those of the industrialized countries would have to be reduced by 80% [13, 18].

In a comparative input-output study for Germany and the United Kingdom, Proops et al. [19] point out that, in addition to CO_2 emissions resulting from their own burning of coal, oil and gas, the individual countries should also be credited with emissions resulting from the production of imported goods and services. Table 2.4 contains concrete figures from recent studies.

Only 15 years ago, scientists who warned of the threats to economic development and social stability posed by changes in the biosphere due to the anthropogenic greenhouse effect were defamed as politicizing alarmists. Only energetic objections by scientific organizations such as the US National Academy of Sciences [20] and

[4] ppm ≡ parts per million.

the accumulation of extreme weather events, interpreted by the public as harbingers of climate change, have brought the attacks to a virtual standstill.

At the United Nations Climate Change Conference in Paris in December 2015, the 197 representatives of the participating countries agreed on reductions in CO_2 emissions, with different reduction targets adapted to the respective level of industrialisation. Following ratification by the parliaments of 55 nations, which are also responsible for 55% of global CO_2 emissions, the Paris Protocol entered into force on 4 November 2016. If the reduction targets are not met, the world will be threatened by mass migrations triggered by climate change. Compared to them the migration flows in the second decade of the twenty-first century, which are described in Sec. 5.1.2, as being caused by wars and corrupt elites and fed by illusions, are only weak precursors.

Europeans north of the Alps could well be among the migrants. After all, the regions of the Earth near the poles are warming up the most because of the anthropogenic greenhouse effect. If this leads to increased freshwater input there, e.g. through ice melt on Greenland and increased rainfall, the Atlantic circulation of cold deep water to the south and warm surface water to the north could come to a standstill. This is because fresh water is lighter than salt water and sinks more slowly than the latter. This weakens the southward deep water current and the northward warm surface water current it drives, the Gulf Stream [21, 22]. Its eastern branch warms Europe. If the system of Atlantic thermohaline circulation becomes unstable and collapses, it will be cold in Europe. In this way, nature would strike back at the continent where industrialisation and greenhouse gas emissions began.

Emissions from an economy based on burning coal, oil and gas are now recognized as a global problem. Its complexity—ocean acidification, feedback-driven approaches to tipping points beyond which drastic, irreversible ecological changes such as the collapse of the Gulf Stream occur—and solar energy alternatives are outlined in [23].

Nevertheless, the president of the USA, who has been in office since the beginning of 2017, questions the anthropogenic greenhouse effect and, as promised in the election campaign and announced on 1 June 2017, wants to withdraw from the Paris Climate Treaty and abolish restrictions on the energy industry that are necessary for climate policy. On the other hand, he wants to introduce new restrictions, namely on trade with the USA and travel to this country. No small part in the election victory of the political outsider is attributed to the target group-specific use of social networks of the Internet. Therefore, constraints not only on the use of energy but also on the handling of information are worth considering. Section 1.2 deals with that.

Heat Barrier

Even without the anthropogenic greenhouse effect, climate problems would occur if energy conversion and entropy production in the biosphere continued to grow.

Calculations of the heat equivalents of pollutants, summarized in Sect. 2.3.2, show that it is in principle possible to convert all polluting emissions of substances, including radioactive ones, by appropriate technical measures into thermal emissions of such quantities that the Second Law according to Eq. (A.4) is satisfied. If one were not afraid of the costs (and risks) involved, one would probably do so, because today thermal emissions are considered to be the most harmless form of entropy production. Heat emissions result from global non-solar energy consumption (about 4.7 10^{20} Ws in 2013), which ultimately ends up more or less in heat.[5]

Even if we do not worry about heat emissions today—unless we have to shut down power plants because of overheating of the rivers from which they take their cooling water—this would change if the world energy consumption per time unit exceeded the 2013 level of 1.55 10^{13} W by a factor of 20. Then the world would face the so-called *heat wall* of about 3 10^{14} W. This is roughly three per thousand of the solar radiant power received by the Earth according to Eq. (A.7). Anthropogenic heat fluxes of this magnitude are likely to change the climate even without the anthropogenic greenhouse effect [24, 25]. Local climate changes, observed in cities where space heating, transport and industry significantly increase heat emissions compared to the surrounding area, already point to the heat barrier.[6]

The heat wall is the final, ultimate limit to industrial economic growth on Earth. Further growth will only be possible to the extent that emissions linked to energy conversion no longer burden the biosphere. We will return to this in Chap. 6.

1.1.3 The Cerberus Before the Land of Milk and Honey

A few years ago, a physicist with a PhD who held a responsible position in a professional association of the physical society of an industrialized country said, "The Second Law of Thermodynamics applies only in closed systems. That is why it cannot be applied to open systems like the economy." Pointing out that the statement of Eq. (1.3), namely "All processes in the real world produce entropy," *is* the Second Law and *always* holds, was to no avail. Unfortunately, his opponent in the discussion did not present the following argument: "If the Second Law does not hold in open systems like the earth and its economy, then build a perpetual motion machine of the second kind. That way you would solve all of humanity's energy and environmental problems and certainly win the Nobel Prize for Physics and probably even the Nobel Prize for Peace."

For the technical formulation of the Second Law of Thermodynamics—the summary of the experience of many failed inventors and of the patent offices to which they submitted and still submit their inventions—reads as follows

[5]The potential energy stored in buildings is relatively small and is released as heat when the buildings are demolished. Similarly, the oxidation of chemical products converts the chemical energy stored in them into heat.

[6]The totality of factors that determine the climate in the "urban islands" is described by [6, pp. 339–343].

There is no perpetual motion machine of the second kind.

A perpetual motion machine of the second kind is defined as a cyclically operating machine that does physical work and in the process does *nothing* other than cool down a heat reservoir.

Even more useful, of course, would be a perpetual motion machine of the first kind. That would be a machine that does work without having to be fed anything from outside. People tried to build such a machine for a long time in vain, about as long as they searched for the philosopher's stone. They failed because of the First Law of Thermodynamics, which states that energy is a constant that can neither be increased nor decreased.

Nevertheless, the promises of new perpetual motion machines are still the basis of new business models today. Some years ago, for example, local authorities in Lower Franconia were offered shares in a Swiss company as a very profitable financial investment. The company promised the production of generators which, as a combination of perpetuum mobiles of the first and second kind, convert the quantum mechanical zero-point oscillations of the vacuum into electrical energy by means of the Casimir force. But a physics student from Würzburg went to Switzerland with the local politicians and successfully prevented them from acquiring the company shares.

If it were possible to build perpetual motion machines of the second kind, it would be possible, for example, to extract part of the heat from the world's oceans, convert it into electrical energy via power plant generators and in this way cover several hundred times the world's current energy requirements without harming the environment.[7]

The inhabitants of a universe which would not be subject to the constraints of the Second Law would live in a land of milk and honey, at least thermodynamically, as soon as they would have reached a degree of industrialization comparable to our present one. For they could use the energy available in their system over and over again for the performance of work in machines without this energy losing any of its usefulness. We, on the other hand, must resign ourselves to the fact that a given quantity of energy loses some or all of its usefulness after the performance of an energy service—be it the lifting of loads, the digging of pits, the propelling of a vehicle, the tempering of rooms, and so on. The useful part of a quantity of energy is called *exergy* the useless part is called *anergy;* anergy is, for example, heat given off to the environment.

About their sum says the First Law of Thermodynamics, the theorem of conservation of energy:

---

[7] According to "World Energy Council Germany 2012", the world energy turnover in 2011 was about 5.4 10^{20} Ws. The energy content of the ocean layer between 0 and 700 m depth is estimated at 10^{23} Ws. Only a very small part of this energy could be used permanently by means of the OTEC (Ocean Thermal Energy Conversion) process. The ultimate source of this energy is the sun, which maintains the temperature difference between the ocean surface and the deep layers [2].

Energy ≡ exergy + anergy = constant.

Energy quantity and the energy quality determined by exergy are discussed in more detail in Appendix A.2.

As a consequence of the Second Law, exergy is destroyed and anergy is increased in every event in our universe. In this sense one speaks of "energy consumption". This is the Cerberus before the land of milk and honey of an unlimited growth of the economy on planet earth.

Nobel Prize-winning economist Robert M. Solow did not quite believe in this Cerberus yet, but if it existed, he feared the worst. So he said in 1974:

> The world can, in effect, get along without natural resources … (but) if real output per unit of resources is effectively bounded – cannot exceed some upper limit of productivity which is not far from where we are now – then catastrophe is unavoidable. [26]

It inevitably follows from the Second Law that with each use of a unit of the "natural resource" energy, the exergy portion contained in it is reduced and finally—often already after the first use—irretrievably lost for the production of the value added, the gross domestic product (GDP), or parts thereof. Whether we are threatened with an economic catastrophe because of this is examined in Chaps. 4 and 6.

1.2 Constraints

> *Confucius* was once asked what he would do first if he were given power in the state. He replied, "Improve the use of language". "Improve the use of language? What is so important about that?" his disciples wondered. Confucius said, "When language is defective, thoughts become confused. When thoughts become confused, science, arts, and law suffer. If science, arts, and law suffer, then the state suffers."

So far, we have been talking about the hard constraints imposed by nature and its laws on the production of material wealth. Looking at recent developments in highly industrialized societies and their interaction with developing and newly industrialized countries, one cannot help but also think about "soft" constraints that citizens should impose on themselves and their communities.

1.2.1 Information Flows

Advances in the micro- and nanostructuring of the transistor, a key element of modern communications technology, have led and continue to lead to a rapid increase in information flows around the globe. For example, "nearly 700 million messages are sent per day … in Germany via WhatsApp … Per minute, users worldwide share 400 h videos on YouTube, 216,000 photos on Facebook, and they like 2.4 million posts on Instagram" [27].

The entropy production associated with information flows due to the heat generated by the electrical currents in the transistors, as well as the consumption of scarce materials, e.g. rare earths, in the production of computers, tablets and mobile phones, will not be discussed further here. But despite—or precisely because of—the appreciation of a well-informed population, might restrictions on information flows not also become necessary?

The previously critically commented encyclical *Laudato Sí* finds clear words for the problem of information overload in its points 46 and 47:

> 46. Social components of global change also include the impact of some technological innovations on work, social exclusion, inequality in the availability and consumption of energy and other services …
>
> 47. Add to this the dynamics of the media and the digital world which, when transformed into an omnipresence, do not favour the development of a capacity for wise living, profound thinking and generous love. In this context, the great sages of the past would risk having their wisdom extinguished amidst the distracting noise of information. This requires us to make an effort so that these media are reflected in a new cultural development of humanity and not in a decay of its innermost richness. The real wisdom that comes from reflection, dialogue and generous encounters between persons is not obtained with a mere accumulation of data that ends up saturating and befogging in a kind of mental pollution. At the same time, there is a tendency to replace real relationships with others, with all the challenges they entail, with a kind of communication mediated by the Internet. This allows us to choose or eliminate relationships at our convenience, and so a new kind of artificial sentiment tends to form that has more to do with gadgets and screens than with people and nature. The current media allow us to transmit and share knowledge and emotions. Yet they also sometimes prevent us from coming into direct contact with the fear, the shudder, the joy of the other and the complexity of their personal experience. Therefore, it should not be surprising that, together with the overwhelming offer of these products, a deep and melancholy dissatisfaction in interpersonal relationships or a harmful loneliness spreads.

"Mental pollution" and the pollution of the biosphere by entropy production have randomness and confusion in common. This is shown by looking at the connection between entropy and information [28], which is presented mathematically in Appendix A.1.3.

Entropy, the physical measure of disorder, and Shannon entropy, the measure of randomness in information theory, differ mathematically only in the probabilities occurring in their definitions and a freely chosen constant. Thus, as explained by Eqs. (A.16), (A.17), (A.18), and (A.19), for entropy one obtains a sum of probabilities each multiplied by its natural logarithms, ln, while for Shannon entropy it is convenient to multiply the probabilities by their logarithms to the base of the number 2 (occurring in the binary processes of digitization).

The elementary entropy unit is the bit. One bit of randomness corresponds, for example, to a coin toss with the same probability for "heads/picture" or "number". In computers, the bit is the unit of information that can be stored in a cell with two possible states. In the transistor, these two states consist of "power on" and "power off". Such states encode the dual numbers "zero" and "one". A bit of entropy is given when 0 and 1 can be encountered in the memory cell with equal probability. In the photosynthesis of CO_2 and H_2O to $C_6 H_{12}O_6$ (glucose) discussed in Sect. 2.1, 40 bits are removed from the biosphere for each glucose molecule formed. When living things produce energy by the respiration of glucose and the emission (and ensuing random distribution) of CO_2 and H_2O, these 40 bits of entropy are returned to the biosphere.[8] In other words, the 40 bits extracted correspond to the loss of degrees of freedom of motion of the atoms as they bind to the fixed sites in the sugar molecule. There they can perform fewer random motions than in the original substances. One can also consider these 40 bits as the information needed to assign the atoms their fixed places in the glucose molecule. In macroscopic systems, typical entropy changes are on the order of 10^{23} bits. This is treated in more detail in [2, 29].

To speak of "mental pollution" through information overload is justified in the following sense. In order to cope with our environmental and social problems, we need to store the (often dry) factual information required for this purpose in our brains in a well-ordered, clear and easily retrievable way. However, due to the abundance of information assailing us, which, especially as images, appeals more and more to emotion and less and less to reason, society is constantly placed in states of excitement corresponding to the highly entropic states of excitation of physical systems, one of which is shown in Fig. 1.1. In this condition, it is difficult to recognize the really important information, to organize it, to store it in a permanently protected way, and to have it present when doing important work and making important decisions. Information overload therefore prevents calm, concentrated learning, consideration and decision-making. And so it enables electoral successes of both sober and charismatic populists, when voters have forgotten the many examples of their failure to address economic and technical issues. In addition, the lingua franca of the Internet is a mixture of languages that often makes senseless connections with Anglo-Saxon advertising slogans and promotes confusion of thought, so that the consequences of faulty language named at the beginning by Confucius are imminent.

Wise restriction of media consumption by children and adolescents is likely to be particularly important. This is pointed out by the Federal Ministry of Health in its "BLIKK Study 2017" published on 29 May 2017 [30], which is based on a survey of 5573 parents and their children. It states:

> The possibilities and opportunities of digitisation are beyond question. But digitisation is not without risk, at least when media consumption gets out of control: the numbers of internet-dependent adolescents and young adults are rising rapidly – experts now estimate that there are around 600,000 internet addicts and 2.5 million problematic users in Germany. With the BLIKK media study presented today, the health risks of excessive media consumption for children are now also becoming increasingly clear. They range from feeding and sleeping disorders in babies to language development disorders in toddlers and concentration disorders in primary school children. If the media consumption of children or parents is

[8] Since all CO_2 molecules emitted by the respiration of food have previously been removed from the atmosphere by photosynthesis, a research project initiated within an international scientific academy to investigate the influence of human respiration on the anthropogenic greenhouse effect was pointless and was abandoned after some back and forth.

> conspicuously high, paediatricians and adolescent doctors detect corresponding abnormalities at a far above-average rate. Marlene Mortler, the Federal Government Commissioner on Narcotic Drugs, commented: "This study is an absolute novelty. It shows what health consequences children can suffer if they are left alone in the digital cosmos to develop their own media competence without the help of parents, educators and paediatricians. For me it is quite clear: We must take the health risks of digitisation seriously! It is urgently necessary to give parents guidance on the subject of media use. Young children do not need a smartphone. They first need to learn to have both feet firmly planted in real life. The bottom line is that it is high time for more digital care – by parents, by schools and educational institutions, but of course also by politicians."

If digital care is inadequate, psychologists will perhaps later talk about the fact that the evolution of "Homo smartphonicus" began in the first decade of the twenty-first century with the market launch of the iPhone. If it is foreseeable that this will lead to a dead end, should we not consider a partial return to the conditions before the establishment of the Internet, when sending and receiving information was paid for with money and not with personal data? Could not taxes be imposed on the operators of social networks in each country, based on the number of accounts that citizens of that country have on Facebook and the like? Could not such a tax also be part of a European Union response to punitive tariffs that the US administration, in office since 2017, wants to impose on European goods?

1.2.2 Capital Flows

At the United Nations climate negotiations, the less industrialised countries insisted and continue to insist on support from the highly industrialised countries. This support should and must be given as help to adapt to the expected climate change—but certainly not in such a way that the aid money disappears into the pockets of kleptocratic elites and ends up in secret accounts of European, North American and Caribbean banks.

Only our discreet bankers and financial advisers have any idea of the billions of dollars that Oriental, Southeast Asian, and African potentates and Latin American landowners extract from their countries and deposit in banks, stocks, and luxury real estate in the industrialized countries and their offshore financial repositories. An inkling is provided by Oxfam's liaison to the African Union in Addis Ababa, Désireé Assogbavi, who said of the idea of a Marshall Plan for Africa: "If our Western partners helped us keep on the continent the US$60 billion of dirty money that flows out of Africa every year, we would be doing just fine, and we could do without any Marshall Plan."[9] And as early as 1970, a normally mild-mannered colleague from the Departamento de Física of the Universidad del Valle in Cali, when asked what would be the fastest way to change the dire social situation of broad sectors of the population in the rich, beautiful country of Colombia, replied: "You'd have to shoot about 30 families." The answer was, of course, purely theoretical and referred to

[9] DER SPIEGEL, 3/2017/43.

"fastest". But it shows the deep bitterness of the population in a country where about 30 family clans dominated the two major parties that had competed for power in the country for 150 years and had fought this competition in bloody civil wars. Section 5.2 goes into this in more detail. While exchanging views on the conditions in industrialised and developing countries, my Colombian colleagues confronted me with the fact that the huge sums of money shipped out of the country by the Latin American elites were gladly received by US, German and Swiss banks. In another South American country, the representative of a leading German automobile company told me how he had to throw the millions in bribes for a major contract to supply fire engines into the wastepaper basket of the interior minister. Until the mid-1990s, German companies were able to deduct the bribes paid to foreign civil servants from their taxes.

The "World Wealth Reports" published by Merrill Lynch and Capgemini show that the *number of* Latin American so-called "High Net Worth Individuals" (HNWI) with financial assets of at least one million US dollars is small compared to the number of all HNWI, but their *share of assets in* the total global assets of all HNWI is in the order of magnitude of the share of European HNWI [2, pp. 229–231], see also "Capgemini and RCB Management World Wealth Report 2013".

And how the rich and super-rich of developing and developed countries unite in an effort to keep the shy deer (financial) capital safe from the tax authorities of their home countries in exotic tax havens has been reported since 2016 by international networks of investigative journalists in the *Panama* and *Paradise* Papers.

Due to the low social progress of the developing countries subjected to the free capital market during half a century, the idea of constraining the global financial capital flows suggests itself. The following procedure may realize it.

Independently of the problem of financial aid to cope with the consequences of the anthropogenic greenhouse effect, the industrially highly developed states, in close cooperation with civil society organizations of the developing and newly industrializing countries as well as reliable international non-governmental organizations, decide on a treaty to introduce a *capital flight drag*. The treaty provides that all banks in the states that are members of the capital-flight-drag agreement are prohibited, under threat of license withdrawal, from accepting funds from natural persons and legal entities from countries with a level of corruption identified by "Transparency International" and classified as unacceptable by the states of the agreement; these funds must not be invested in whatever form, outside the home country of the person/entity concerned. Trade between corrupt countries and the rest of the world would be conducted through United Nations transfer banks. Their personnel should be selected, in contact with non-governmental organizations and state institutions of international cooperation, from the economically competent representatives of domestic civil societies who are willing to work for salaries of, and as diligently as, senior administrators of the states with low levels of corruption.

The capital flight drag is likely to be one of the few instruments with which the industrially highly developed countries can help to create the necessary conditions for an improvement of living conditions in the poorer countries of the "South" without violating the peoples' right to self-determination. It would deny their ruling, corrupt elites, who cling to power and thereby also trigger tribal feuds and civil wars, access to the international financial markets. Then these elites would have to invest their spoils made in governing their own country—in other words, they would have to use the financial capital now remaining in the country to purchase real capital, i.e. machinery and the buildings and equipment needed to protect and operate it, and to develop efficient, sustainable agriculture. And just as the transition to industrial society in European feudal agrarian society found enlightened members of the ruling class who cared about the social and technical progress of their countries, so today in the developing world, when the money drainage channels are plugged, smart men and women are likely to come to power who do not sell off the often rich natural resources of their homelands to foreign countries, but use them to educate their fellow citizens and create industrial jobs. In the short term, this will exacerbate the emissions problem. In the long term, everyone must work together to find solutions.

Restrictions on the movement of capital by imposing sanctions on banks and companies in industrialized countries for violations are by no means new. They were introduced by the USA and the European Union once against Iran because of its nuclear programme and another time against several collaborators of the Russian President Putin because of the annexation of the Crimea.[10] Restrictions on the capital markets are thus a well-tried instrument for the enforcement of higher interests. And eliminating the worst forms of personal corruption from twenty-first century governance should be in the highest interest of the international community and industrially advanced countries.

Irrespective of this, there is another reason to introduce restrictions on the movement of capital.

Economists Wolfgang Eichhorn and Dirk Solte [31] point out in their book *Das Kartenhaus Weltfinanzsystem (The World Financial System: A House of Cards)* that the global financial system would collapse if all holders of savings accounts and securities demanded that their balances be paid out in central bank money, the only legal tender, because: "Compared to the central bank money in circulation, there are … more than fifty times as many securitized money claims." These securitized money claims are traded internationally, with transactions occurring at the speed of light thanks to modern information technologies. They are causing capital flows around the globe to swell stronger and faster, with the risk of instabilities occurring. Thus, from August 2007, a severe financial crisis developed, partly because of the trading of bad loans on US real estate. The US investment bank Lehman Brothers collapsed on 15 September 2008. The bursting of the real estate bubble also caused distress for banks in other countries. Germany had to bail out Hypo Real Estate with taxpayers' money. This led to the first major global economic crisis of the twenty-first century. Whether the subsequently tightened legal regulations for the banking

[10] Crimea came under Russian rule in 1783. Under Soviet party leader Nikita Khrushchev, it was given to the Ukrainian Soviet Socialist Republic in 1954. After the collapse of the Soviet Union, Ukraine became a sovereign state. In 2014, Crimea seceded from Ukraine and was de facto annexed to Russia.

sector will be sufficient to prevent a repeat of something similar is not certain. After all, the economic players had apparently learned nothing from the bursting of the Japanese real estate bubble in the early 1990s. That bubble was inflated by speculation that led to the Tokyo Imperial Palace and its gardens being valued as high as the entire state of California [32].

1.2.3 Markets

Systematic thinking about the origin and distribution of wealth began with the Scottish moral philosopher and economist *Adam Smith* and his book *An inquiry into the nature and causes of the wealth of nations*. This work systematically describes the liberal economic doctrines of the eighteenth century and became the bible of classical national economics. It was published in 1776, the year of the Declaration of Independence of the United States of America.[11] "This coincidence in time is hardly accidental: there is a connection between the political struggle against the form of rule of the monarchy and the emancipation of free market pricing from state intervention and regulation," notes the author of one of the most successful economics textbooks [33] *Paul A. Samuelson*.

This expresses the basic concern of economics, which today, more than any other scientific discipline, determines the lives of individuals and relations between peoples: It is about the formation of the prices of goods and services on free markets. According to Adam Smith, an "invisible hand" ensures that individuals, in the pursuit of their self-interest, do precisely what serves the good of the whole. Nevertheless, Adam Smith by no means rejected from the outset all economic policy interventions by the state and the resulting restrictions on markets. This was not the case until the later Manchester capitalism, which, as an extreme form of economic liberalism, advocated unconditional free trade and unrestricted economic freedom and is also referred to as "laissez-faire capitalism".

In the 1980s, U.S. President Ronald Reagan and British Prime Minister Margaret Thatcher rediscovered the appeal of Manchester capitalism and revived this radical version of the market economy, now called Reaganomics or Thatcherism, which had also been propagated in continental Europe in the name of freedom.

This revival is all the more astonishing because after the Second World War the Western industrialised countries, not least Germany, in accordance with their self-image as social constitutional states, did not want to rely solely on the "invisible hand", but gave the market a legal framework within which the individual striving for profit could develop in such a way that economic progress and growing prosperity for all went hand in hand.

The regression initiated by Reagan and Thatcher was favoured by a profound change in the modes of production to which the framework conditions of the social

[11] In the same year, the (precursor of the) steam engine developed by the Scotsman *James Watt* Newcomen's steam pump was also used commercially for the first time to pump coal mines dry.

market economy have not yet been adapted. This change consists in the advance of information technologies, triggered by the invention of the transistor and sometimes understood as the second (or even third) industrial revolution. The combination of heat engines and transistors has given new, unimagined impetus to automation. More recently, "digitization" has come to mean automation. It is what determines our economic and social destiny today. Chapter 3 is about it. It is true that Reaganomics and Thatcherism are in retreat. The first great stock market crash of the twenty-first century and the subsequent global recession, to which their deregulatory ideology gave rise, saw to that. But they have damaged the reputation and acceptance of the market economy.

As a result, and also in view of the ecological problems, alternatives to "capitalism" with restrictions on production, trade and markets are now being demanded by very different political movements and schools of economic thought—and this at a time when most countries in the world are striving for industrialisation under market-economy conditions.

Against the background of the intensifying critique of capitalism, it should already be said here that in this book the production factor capital is defined in Chap. 3 as consisting of all energy conversion devices and information processors together with the buildings and installations required for their protection and operation. This physical capital of the real economy is activated by energy. The monetary capital of the financial economy is quite different: it is activated by human endeavours. These can lead to investments in physical capital. But they can also inflate speculative bubbles whose bursting has already caused great damage to the real economy.

In this chapter we have considered the currently perceptible disruption of industrial society and the constraints required for its control. The following chapters deal with the thermodynamic foundations of industrial value creation, quantitative analysis of economic growth taking into account energy as a factor of production, and the guard rails for development paths on which a Dark Age can perhaps be avoided.

2 Energy and Life

> Power of fire, how beneficial
> if carefully guarded and harnessed by man.
> Whatever he forms, and what he creates,
> he owes it to you, o gift of the gods. (Friedrich von Schiller, "*The Song of the Bell*")

The taming of fire in the darkness of history 400,000 ago was man's first step on a path that has led him high above all other living beings on earth: With at first slowly and then vehemently unfolding technical creativity, he used and still uses the forces of nature more and more for the increase of wealth and mastery of the world.

Prometheus, who according to Greek myth brought fire from Olympus to earth, was cruelly punished by Zeus, the father of the gods, for giving such power to humans and thus saving them from doom.

Fire gives light and warmth for home and hearth, metalworking, material transformation and work performance. Saints and poets have sung of the power of fire.[1] The source of the power of fire remained hidden from man for a long time. In ancient natural philosophy, fire was considered the origin of being, or one of the four elements.

It was not until the nineteenth century that physicists learned, not least from the German physician *Robert Mayer* and the English brewer *James Prescott Joule,* that a quantity appears in fire which is inherent in wood and coal, food, location and movement and which can be converted into heat and work. Today we know that this quantity gave rise to our universe in the Big Bang some 14 billion years ago, and that it sustains all life on earth through sunlight and the photosynthesis of plants. Its

[1] St. Francis of Assisi prays in his Canticle of the Sun [34], "Praise be my Lord for brother fire/ through which you illuminate the night/His spraying is bold, serene he is, beautiful and mighty strong."

R. Kümmel et al., *Energy, entropy, creativity*,
https://doi.org/10.1007/978-3-662-65778-2_2

name, *energy*—understood as stated in Sect. 1.1.1—was first pronounced by *Thomas Young* (1773–1829). Energy and its transformation into work is essential for all the life and the economic activities that have created the biosphere and developed the civilizations on our planet.

An extraterrestrial observer who has followed the development of life on Earth for 4 billion years would be able to determine only one quantity as the driving force: the irradiated solar energy. It is the factor that, together with genetic information processing, produces everything on earth. That is why the sun enjoyed divine veneration in many religions.[2]

2.1 Photosynthesis and Respiration

Photosynthesis provides the chemical energy for life on Earth. Respiration converts this energy into the work done in and by living organisms.

Photons from sunlight excite electrons in the chlorophyll molecules of the photosynthetic reaction centers of plants, algae and some bacteria. The excited electrons flow along a molecular chain as a tiny, solar-driven electric current. This current does two jobs. It splits water molecules into hydrogen and oxygen atoms, and it converts adenosine diphosphate (ADP) molecules into higher-energy adenosine triphosphate (ATP) compounds. In a series of chemical reactions, sugar (glucose) and oxygen are formed from hydrogen, carbon dioxide, and ATP. In the photosynthetic summation equation, six molecules of water and six molecules of carbon dioxide yield one sugar molecule and six molecules of oxygen when exposed to light:

$$6H_2O + 6CO_2 + \text{Light} \rightarrow C_6H_{12}O_6 + 6O_2.$$

In this process, the energy of the sunlight is essentially converted into the chemical energy of sugar and oxygen and thus stored.

The respiration process completes the fundamental life cycle. In this process, the stored solar energy is released by plants and animals through the recombination of sugar and oxygen to water and carbon dioxide in order to perform work with it, e.g. mechanical work of muscle contraction, electrical work when charges are transported, osmotic work when material is transported through semi-permeable partition walls, or chemical work when new materials are synthesized. In these nonequilibrium processes, work output can only be accomplished if some of the stored exergy is converted to heat and released into the environment, where it ends up as anergy. An essential mechanism of this entropy disposal is water evaporation. That is why plants are so thirsty.

[2]The Egyptian pharaoh Akhenaten (Amenophis IV) hymnally praised the sun around 1400 B.C. [35]: "You appear beautiful in the mountain of light in the sky,/Living star of the sun, you who lived in the beginning … ./You fill every land with your beauty./You are great, sparkling above every land,/Every land embraces your rays/until the last end of everything created by you".

In the sum equation of the respiration process, which in a series of oxidation and reduction reactions converts the exergy stored in the sugar into the chemical energy of ATP,

$$C_6H_{12}O_6 + 6O_2 \rightarrow 6H_2O + 6CO_2 + \text{“38ATP”},$$

“38 ATP” stands for the energy of about 2800 kJ stored in 38 moles of ATP. ATP serves as the universal energy currency for all living things. If work must be done at a given time, ATP is converted to ADP and inorganic phosphate and its energy is released in a hydrolysis reaction. The chemical end products of respiration are water and carbon dioxide emitted into the atmosphere.

Since the “Neolithic Revolution” after the end of the last ice age, agriculture has systematically put photosynthesis at the service of man in a climate that has become stable. Arable farming systematically produces and breeds cereals, tubers, fruit and vegetables, and livestock farming converts plants that are inedible for humans into nutrient-rich meat.

2.2 Agricultural Society

The use of solar energy and the energy reservoirs created by it have determined the history of human civilization. Thus, from today’s perspective

> the universal history of mankind ... can be divided into three periods, each of which corresponds to a specific energy system. This energy system sets the framework conditions under which social, economic or cultural structures can form. Energy is therefore not only one factor among others, but it is in principle possible to determine the formal basic features of the corresponding societies from the respective energetic system conditions. [36]

The first of these three periods was the age of hunter-gatherers. It covers 99% of the time of human existence. This was followed by the age of farmers and craftsmen, which lasted about 10,000 years. It was succeeded 200 years ago by the present industrial age. The progress of civilization in the course of history went hand in hand with the increasing use of energy in newly developing areas and the development of ever more productive sources of energy.

Thus, for the longest period of its existence, humans met their energy needs in the form of food and firewood directly from solar energy fluxes and the biomass they produced. The energy consumption per capita and day was:[3]

- 2 kWh for the simple gatherer of plant food during the 600,000 years before the taming of fire;
- 6 kWh for the hunter-gatherer with a domestic hearth 100,000 years ago;

[3] For the sake of simplicity, the kilowatt hour (kWh) familiar from everyday life is used as the unit of energy. 1 kWh = 3,600,000 Ws, 1 Ws = 1 J.

- 14 kWh for the simple farmer who developed agriculture and animal husbandry in the Neolithic revolution after the beginning of the current, climate-stable warm period 10 to 12,000 years ago;
- 30 kWh for members of highly civilized medieval agrarian societies in 1400 AD, in which craftsmen made increasing technical use of fire, especially for metalworking [37].

This has to be seen alongside the following estimates of the annual energy yields per hectare of different agricultural production methods (including fallow) [36, 38]:

Rice with fire clearance (Iban, Borneo)	236 kWh/ha
Horticulture (Papua, New Guinea)	386 kWh/ha
Wheat (India)	3111 kWh/ha
Corn (Mexico)	8167 kWh/ha
Intensive farming (China)	78,056 kWh/ha

Hunter-gatherers, on the other hand, achieved only 0.2–1.7 kWh per ha and year. Thus, on the basis of Chinese intensive farming 50,000 times more people could live on a given area than under hunter-gatherer conditions.

After the transition to an industrial society through the development and use of fossil energy sources, energy consumption per capita per day in the industrialized countries has been increasing much faster than in the rest of the world since the nineteenth century. It was in the year:

- 1900 in Germany: 89 kWh,
- 1960 in (West) Germany 61 kWh and in the USA: 165 kWh (energy was used more efficiently in 1960 than in 1900),
- 1990 in (West) Germany: 117 kWh and in the USA: 228 kWh,
- 1995 in reunified Germany: 133 kWh, in the USA 270 kWh; in the world average: 46 kWh, in the developing countries: 20 kWh.

In 2011, the world average daily consumption of each person was 59 kWh.

In the agrarian society, the soil as a carrier of the plants that collect solar energy was the source of economic and political power. Working the soil, as well as the wood, stones and metals extracted from it, only human and animal muscle power were available. The advanced agrarian civilizations that shaped human civilization for more than 5000 years and stretched from Mesopotamia of the Euphrates and Tigris to the Far East, encompassed the Mediterranean region and Europe to the west, and also left behind impressive testimonies of art and architecture in Central and South America, were limited in their development by the limitation of the amount of work that could be done from muscular power and the low efficiency of energy conversion in humans and animals. Inclined planes, pulleys, levers, windmills, and watermills offered some, though very limited, help in overcoming the biological barriers. An idea of the power achievable by muscular strength is given by the following figures [39].

The pulling force of a horse is 14% of its body weight and is about 80 kilopond (kp) permanently (1 kp = 9.81 N). Deep ploughing requires 120–170 kp and mowing 80–100 kp. The average power of a horse is 600–700 W, that of a donkey 400 W. A horse performs a work of 3–6 kWh/day. For this it needs feed with an energy content of about 30 kWh. Its energy efficiency is therefore between 10% and 20%. For certain mechanical work, donkeys were used. As a rule, a maximum of four donkeys could be harnessed to a whim, so that its output was limited to a maximum of 2 kW. The energetic limits of overland transport in agrarian society were given by the fact that a horse would eat a wagonload of fodder in a week. It was therefore pointless to use horse and cart to transport food for longer than a week. Therefore, during longer campaigns, armies had to feed themselves from the lands they occupied (and "devastated").

> The energetic efficiencies of humans and horses are similar. However, the average power of a human being is only 50–100 W and thus only one eighth of the power of a horse. In one aspect man was and is superior to any animal, however strong: In the intelligent control of the energies released in his body upon the objects of the external world for the useful transformation of the same. Thus man's hand, in combination with the processing of information in the human brain and the mysterious power we call creativity, has accomplished the highest cultural and technical feats. These were, however, limited by the human maximum power of 100 watts. That is why the advanced agrarian civilizations needed masses of disenfranchised slaves, serfs and bondmen to build their pyramids and temples, castles and fortresses and, most importantly, to cultivate their lands, which enabled their masters to live a life worthy of a human being, even according to today's standards. [38]

When the apostle Paul wrote the letter to Philomenon asking for clemency for the slave Onesimus, the population of the Roman Empire was 25% slaves. The Roman Empire collapsed under the onslaught of Germanic barbarian tribes, civilizing them nevertheless in its agony. New feudal societies arose in Europe beyond the Alps. There, from the early Middle Ages onwards, the originally free peasants became increasingly dependent on the noble feudal lords and landowners. They became bondmen and serfs. The niceties of courtly life were reserved for the princes, the artists they patronized, and the merchant financiers of the court.

After the discovery of America, almost ten million enslaved Africans were shipped to the New World between 1520 and 1850 to work enslaved on the plantations. Their owners lived a cultivated and splendid life.

Slavery and serfdom were the preconditions for the high cultural achievements of agrarian societies. The splendour of the few outshone the misery of the many.

Even after the abolition of slavery and serfdom in the countries encompassed by the Enlightenment during the eighteenth and nineteenth centuries, power derived from the ownership of land continued to determine social rank and political influence for several generations. With the advance of industrialization, land ceded this significance to capital.

2.3 Industrial Society

The transition from an agrarian society to an industrial one coincided with technological advances since the end of the Middle Ages and the development of empirically based natural science. From them sprang the heat engines, as well as all other energy conversion equipment and information processors that form the productive capital. The educational system has been constantly adapted to the requirements of the industrial society's development dynamics.

In the struggle for the rules of economic policy according to which the forces of nature are put at the service of man, mercantilism, whose goal was the financial strengthening of state power, and in the nineteenth and twentieth centuries planned economic ideas influenced by Marxism fought against the variously developed free-market doctrines. According to the latter, production and distribution of agricultural and industrial goods and services should be carried out by private enterprises interacting via supply and demand. In this context, in a market economy, the role of the state is usually limited to creating the legal framework appropriate for the production and distribution of the "wealth of nations" and to enforcing its observation; occasional, controlling state intervention in economic activity is not excluded. *Samuelson* characterizes this economic order as an "economic mixed system in which both public and private institutions influence and determine the economic process" [33, Vol. I, p. 65]. In this system, goals such as social justice and sustainability have more of a place than in the previously mentioned "laissez-faire capitalism".

From a cybernetic point of view, the market economy has the advantage over the planned economy that the decision-making processes between the market participants take place quickly in short-circuited control loops, whereas in a planned economy they have to pass through all the successively connected controllers of the planning bureaucracy. This advantage has led to the fact that in the second decade of the twenty first century almost all economies in the world, with the exception of North Korea and Venezuela, operate in a market economy.

The "bible" of national economics, Adam Smith's *Wealth* of *Nations*, was published in 1776. This is the year in which the Declaration of Independence *of the* United States of America created the ideational basis for bourgeois democracy and in which James Watt's steam engine, which opened up the industrial age, found its first commercial application in pumping coal mines dry. As a man of the agricultural age that was just ending, and without being able to recognize the completely new thing that had come into the world with the coal-fired steam engine, Adam Smith emphasized human labor, combined with land and the agricultural means of production capital,[4] as the source of wealth. On the whole the productivity

[4]In agrarian society, "capital" consisted essentially of hearths, tools, foundries, vehicles and implements moved by muscle power, sometimes by water and wind, and the buildings that housed them. The Latin word for money, *pecunia*, originally meant ownership of livestock. How far firearms were counted as "capital" is unclear.

of labour is increased by the division of labour, which can and should also take place on an international level by means of free trade.[5]

But without heat engines in Adam Smith's economic world, the realization of the human rights to life, liberty and the pursuit of happiness proclaimed by the American Declaration of Independence would hardly have been possible in today's, albeit still unfinished, form. It is they, after all, that convert the fuel reserves laid down by nature on earth, into enormous energy services and largely free the people of the industrialized countries from hard physical labor as their prosperity grows.

Heat engines, the first of which, the steam engine, is now obsolete, convert the heat extracted from (chemical and nuclear) fuels into work; the conversion processes are controlled by information processors such as valves, relays, electron tubes, and transistors. The heat engines, together with furnaces and reactors, form the heart of the real capital stock. They are (still) indispensable for the production of the goods and services of industrial economies. Important today are:

Steam turbines. They power steamships and electricity generators of power stations; through the latter they operate tramways and railways and also all electrical machines in industry and trade such as welding robots, milling machines, drills, computers, printers, and household appliances such as kitchen ranges, refrigerators, washing machines, vacuum cleaners and televisions.[6]

Gasoline engines power passenger cars, aircraft, and boats.

Diesel engines power the tractors and harvesters that mechanized agriculture, as well as construction machinery, passenger cars, trucks, locomotives, ships, and decentralized electricity generators.

Gas turbines perform mechanical work via the drive shafts of helicopters, ships, pumping stations, and gas turbine power plants. As jet engines, they generate the thrust for the world's aviation fleets.

While the heat engines powered by coal, oil, gas and some nuclear energy freed and still free man from physical work, the information processors powered by electrical energy increasingly relieve him of routine mental work as well. Essential to this today are transistors, the first examples of which were developed between 1946 and 1948 by *John Bardeen, Walter Brattain* and *William Shockley,* who received the Nobel Prize for this in 1956.

---

[5] In contrast, Karl Marx's labour theory of value sees human labour as the primary source of all wealth. Therefore, the factor labour is also entitled to the entire value added, including capital goods.

[6] An 800 MW steam turbine illustrated in the *Großer Brockhaus* has a length of 44 m and a width of 14 m including the generator. A horse with an output of 700 W needs a pasture area of about 10,000 m^2. All the horses together, which could produce the power of an 800 MW steam turbine, would occupy an area of almost 11,500 km^2. That is about half of the area of Hesse that would have to be converted into pasture land so that the biomass growing on it could, in purely mathematical terms, provide the same energy services via horse bodies as the steam turbine of a power plant driven by coal combustion or nuclear fission.

2.3.1 Technological Progress, Energy Slaves and Emissions

Industrial development is proceeding in bursts of innovation that are opening up new and further fields for the use of energy. The trend has prevailed for 200 years:

In the course of technological progress, energy slaves instead of human beings are increasingly used for the production of goods and services.

Here, an energy slave works in an energy conversion plant and has a primary energy demand of just under 3 kWh/day, which corresponds to the work calorie demand of a heavy worker.

Energy slaves represent *exergy*. As stated in Sect. 1.1 and elaborated in Appendix A.2, exergy is the fraction of a quantity of energy that can be fully converted into work, whether mechanical, electrical, chemical, or any other form of work. The primary energy sources of coal, oil, gas, nuclear fuels, and sunlight are essentially 100% exergy. *Exergy* is *the* central term used in modern engineering thermodynamics to denote the quality and value of quantities of energy.

One can imagine dynamite cartridges as the simplest energy slaves. While thousands of enslaved prisoners of war drudged their souls out of their bodies in ancient quarries, and even later, under more humane working conditions, breaking stone was one of the most difficult and dangerous physical tasks, since Alfred Nobel's invention a few workers have been sufficient to drill holes in quarry walls, stuff dynamite cartridges into them and ignite them. Less noisy and steadier are the energy slaves made up of the fuel-air mixture of an internal combustion engine, which move pistons, connecting rods and drive shafts with each ignition. Perhaps most elegantly, the current- and voltage-derived energy slaves do their work in electric motors and semiconductor transistors.

Wherever we look in our technical-industrial world, we see energy slaves at work, relieving man of dangerous, difficult, or just routine work, enabling him to undertake things he could never do on his own, such as flying.

If one considers a given economic system and asks quantitatively for the number of energy slaves working in it, one obtains this number from the average daily primary energy consumption of the economic system divided by the human work calorie requirement of 2500 kcal (2.9 kWh)/day for heavy physical work. Using this method of calculation, and taking into account the energy consumption per capita per day given in Sect. 2.2, the numbers of energy slaves (ES) serving each person in the various societies throughout history are approximately:

100,000 years ago, hunterers and gatherers using fire			1 ES,
7000 years ago, simple agrarian societies			4 ES,
AD 1400, medieval Western Europe			9 ES,
AD 1900, Germany			30 ES,
AD 1960, (West) Germany:	21 ES,	USA: 59 ES,	
AD 1990, (West) Germany:	40 ES,	USA: 79 ES,	
AD 1995, Germany:	45 ES,	USA: 92 ES,	
World average:			15 ES,
Developing countries:			6 ES,
AD 2011, World average:			19 ES.

An idea of the efficiency and cost of an energy slave is given by the following simple example: Let us assume that an energy slave, working in an electric motor driven by household electricity, is to lift by an elevator a mountaineer weighing, together with his equipment, 100 kg from sea level to the summit of the 8848 m high Mount Everest. To do this, it would have to perform a lifting work of 2.41 kWh. If we assume that the same work has to be done again to overcome frictional resistance and that the price of a kilowatt hour of electrical energy is 20 cents, the climber will be hoisted to the top of Mount Everest at an energy cost of about 0.96 EUR.[7] The total cost would, of course, be increased by the proportionate investment cost of the elevator. But the mass transport of tourists and their equipment to high mountain regions by lifts, cable cars and helicopters shows how much cheaper and more readily available the lifting work of energy slaves is than that of humans.

The energy *quantity* contained in a fossil or nuclear energy carrier, which is the basis for the calculation of the energy slave numbers, are measured by the heat Q released to the environment in a chemical or nuclear reaction under precisely controlled conditions. Table 2.1 shows the energy quantities, determined by their average calorific values, of primary energy carriers of given masses or volumes. More details are given in Annex A.2.

The energy yields of carbon and hydrogen combustion are given by the reaction equations

$$C + O_2 \rightarrow CO_2 + 394\ \mathrm{kJ/mol},$$

$$2H_2 + O_2 \rightarrow 2H_2O + 242\ \mathrm{kJ/mol}.$$

The percentage energy gain from carbon combustion is about 80/70/60/50% for lignite/hard coal/oil/natural gas. Hydrogen combustion provides the rest. Viewed in

[7]Transport mass $m = 100$ kg, acceleration due to gravity $g = 9.81\ \mathrm{m/s^2}$, lifting height $h = 8848$ m, lifting work $mgh = 2.41$ kWh. (To do twice this work, an energy slave has to work for about 1.7 days).

Table 2.1 Mean specific calorific values of primary energy sources in kilojoules per kilogram (kJ/kg) and kilojoules per cubic metre (kJ/m^3); 1000 kJ = 0.278 kWh

Oil	41,900 kJ/kg
Hard coal	29,300 kJ/kg
Lignite	8200 kJ/kg
Wood	14,650 kJ/kg
Peat	12,600 kJ/kg
Oil gas	40,730 kJ/m^3
Natural gas	32,230 kJ/m^3

this way, natural gas appears to be the least climate-damaging fossil fuel. However, its extraction in Siberia and transport through pipelines to Central Europe, as well as its production through fracking in the USA, releases not small amounts of methane. As already mentioned, its molecular global warming potential is 25 times greater than that of carbon dioxide. Whether and how problematic this will make the replacement of coal-fired power plants with gas-fired power plants is not yet clear.

The complete conversion of the mass $m = 1$ g into energy according to $E = mc^2$, where c is the vacuum light velocity, results in 25 million kWh. The same amount of energy is also obtained from the combustion of 3100 t of hard coal or the 19 h operation of a 1300 $MW_{electric}$ boiling water reactor.

The energy slaves operate incinerators, heat engines, and transistors. In the process, they perform work and process information. In this way, they create a large part of the economic value added, the totality of which in an economy forms the gross domestic product (GDP). It is the sum of all goods and services, valued at their market prices, produced by the economy in a given period, e.g. a year. As long as the economic value added was distributed to broad sections of the population in accordance with productivity progress via collective agreements and social transfers, the general prosperity and thus the acceptance of the social and economic system grew in all groups of society.

In 1991, for example, according to Table 2.2, the average working time of a German industrial worker whose remuneration had to be spent on the purchase of basic goods for daily use was, for many goods, less than one third of the corresponding working time in 1960.

At the same time, the inflation-adjusted gross domestic product of the old FRG more than doubled between 1960 and 1989, and in reunified Germany it rose by a further 16% between 1990 and 2000 (Fig. 3.2).

That is why the social market economy system became so attractive to people in East Germany (GDR) and other countries with socialist planned economies that they tore down the Berlin Wall, tore down the Iron Curtain, and joined the European Union.

If *Karl Marx*, when he published *Das Kapital* in 1867, had recognized the completely new thing that had come into the world with the steam engine, he would have seen that surplus value in the sphere of production could be generated by exploiting energy sources instead of exploiting people. Society would have been spared the theory of the pauperization of the masses under capitalism and its inevitable collapse, and instead of suffering the failed attempt to establish a dictatorship of the proletariat, the people of the former socialist

Table 2.2 Purchasing power of the wage minute in (West) Germany in 1960, 1991 and 2008

Good	1960	1991	2008
Bread (1 kg)	20	11	11
Butter (250 g)	39	6	5
Sugar (1 kg)	30	6	5
Milk (1 L)	11	4	4
Beef (1 kg)	124	32	32
Potatoes (2.5 kg)	17	10	14
Beer (0.5 L)	15	3	3
Petrol (1 L)	14	4	6
Electricity (200 kWh)	607	191	201
Fridge	9390	1827	1432
Washing machine	13,470	3207	2008

The average working time of an industrial worker, the remuneration of which were to be spent in acquiring the above goods, is given in minutes [40]

countries, like their more fortunate contemporaries in free-market democracies, would have been able to participate in the surplus value creatively created from energy sources. The tragic failure of socialism to understand the industrial production process is symbolically demonstrated by the fact that the hammer and sickle, the tools of the craftsmen and peasants of the past agricultural epoch, had adorned the state flag of the second most powerful industrial nation on earth on its way to economic collapse. [38]

The Cold War, which could have led to the self-destruction of mankind at any time, ended with the dissolution of the Soviet Union. The horror vision that haunted everyone in the 1980s who was aware that the Soviet Union, as a military colossus on economic feet of clay, was marching towards economic collapse, has faded: despite NATO, the tank armies of the Warsaw Pact would break through the North German Plain and the Fulda Gap into Western Europe. Thus the Soviet empire, with its conventionally superior forces, would grab the fleshpots of the European Community in order to feed on them for a while longer. A kind fortune and *Mikhail Gorbachev* prevented that.

If one sees the peaceful outcome of the Cold War as a sign that the world was not created to perish from the stupidity of mankind, there is reason to hope that it will not come to an extreme in the future—even if the victorious capitalist system now threatens to degenerate on its part and charge the world with new conflicts.

The conflicts are sparked by the growing inequality of life chances within nations and between states and by the restrictions mentioned in Sect. 1.1, which are imposed by the first two laws of thermodynamics on a further growing production of material wealth.

In summary, the First and Second Laws of Thermodynamics say:

Nothing can happen in the world without energy conversion and entropy production.

Energy conversion moves the world, and the entropy production inseparably linked with it leads to energy depreciation and the emission of heat and material flows. No law of nature is more powerful than the thermodynamic laws to which all developments in the universe are subject. Any theory that violates them will inevitably end up on the rubbish heap of science, and any economic system that ignores them will fail. The importance of thermodynamics for economics is also emphatically pointed out by Robert U. Ayres [41, 42].

Nevertheless, the central, decisive role of energy for economic practice is not as clear to many as it is to most engineers. Even some physicists only grant it a kind of "lubricant function". And famous economists explain their belief in its—and all economic goods'—substitutability. At an international conference on natural resources, for example, a young economist pointed out in a lecture that, because of the Second Law of Thermodynamics, energy cannot be substituted by capital at will. A highly respected American economist angrily interrupted him with the words: "You must never say that! There is always a way for substitution."

But there is no substitute for energy. Technical measures can be taken to increase the efficiency of energy conversion systems and thus reduce the energy required for energy services such as the heating of rooms by stoves, the performance of mechanical work by heat engines and the processing of information by electronic devices. However, the first two laws of thermodynamics impose physical limits on this, as discussed in Sect. 1.1. In Sect. 2.3.3 we look at examples of what is achievable.

When all potentials for energy conservation have been exhausted and the energy demand of mankind continues to grow—on the one hand because of the high pent-up demand of newly industrializing and developing countries, and on the other hand because Homo sapiens will still beget many more of his kind—energy sources will have to be developed that are more productive and environmentally compatible than the fossil energy sources dominating Fig. 2.1. If we consider not only their known *reserves,* but also their estimated *resources*, which would have to be extracted at higher costs than the reserves using more advanced technologies, we arrive at a depletion period of about 1000 years for coal at extraction rates of the year 2005 [2]. For the time being, the fossil energy system has its problem (only) with the Second Law of Thermodynamics.

But the problem is so great that humanity must switch as soon as possible to non-fossil energy sources, i.e. those that convert mass directly into energy. This is done by nuclear fission and nuclear fusion. Germany decided in 2011 to abandon nuclear fission. Now, energy of solar origin is to quickly take the place of the previous energy sources.

The sun shines thanks to nuclear fusion. By the fusion of 600 million tonnes of hydrogen into helium, it converts 4.3 million tonnes of matter per second into radiant energy, of which a fraction, namely $1.7\ 10^{17}$ Ws, hits the earth's atmosphere every second (Appendix A). This is about 10,000 times the current global commercial energy use according to Fig. 2.1. There is no doubt that in the future the radiant energy supplied free of charge by the fusion reactor sun can and must be used for the environmentally friendly production of material wealth.

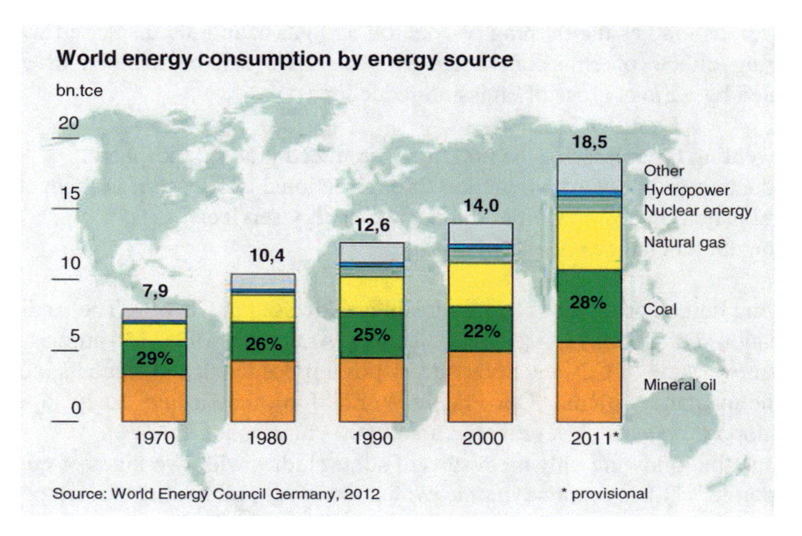

Fig. 2.1 World energy consumption by energy sources [43]. 18.5 billion tce/year = 1.72 10^{13} W

However, the conversion of solar energy into work and heat services is by no means free. Rather, it requires investment in solar thermal, photovoltaic, biomass extraction and processing, wind turbines, dams, energy distribution networks and energy storage facilities.

A hopeful international approach to the use of solar energy in the countries of the EUMENA region (EUrope, Middle East, North Africa) was the DESERTEC project. Power plants in the sunny countries around the Mediterranean were to produce electricity for these countries and Central Europe. Extra-high voltage direct current lines would be suitable for transporting electricity over long distances. On 30 October 2009, an agreement was concluded between the DESERTEC Foundation and the DESERTEC Industrial Initiative "DII GmbH" in Munich. Many large German companies such as Siemens, Munich Re and RWE were part of the initiative. Investments of up to 400 billion EUR were planned. The project status in 2010 is described in [2, pp. 83–85]. Unfortunately, the political upheavals following the outbreak of the Arab Spring did not promote greater progress by DESERTEC. The state of affairs in 2016 and an overview of the wealth of possibilities and plans for the use of renewable energies is given in [23].

Now it is easier to show what is technically feasible in the energy sector than to calculate the associated system costs. Examples of cost-benefit analyses of the integration of renewable energy sources into the European energy system are provided by the publication of the European research project of the LIT Research Group [44]. Perhaps it will also be possible within the next 50 years to bring solar fire to earth in nuclear fusion reactors; nuclear fusion in hydrogen bomb explosions will hopefully remain a thing of the past.

If solar energy is used intensively, by far the largest part of the associated entropy production remains in space and does not burden the Earth's sensitive biosphere.

However, as long as the burning of coal, oil and gas cannot be dispensed with, the damaging effects of emissions listed in Sect. 1.1.2 can be avoided or at least mitigated by various types of emission reduction:

1. Preventing the release of the pollutants produced into the biosphere,
2. Reduction of pollutant production through rational energy use, i.e. reduction of combustion processes with undiminished energy service,
3. Behavioural changes of energy consumers.

Changing human behaviour is difficult and is addressed in Chap. 4. The studies and publications on options (1) and (2) fill libraries. As early as 1995, 134 articles in [45] gave an overview of CO_2 retention and disposal procedures, and since its inception with the inaugural volume "Our Fragile World" [46], the on-line "Encyclopedia of Life Support Systems" has grown to about 600 volumes [47].

But in the following only the results of some studies with two kinds of questions are reported: (1) Is a thermodynamic evaluation of all pollutant emissions possible, and is it suitable to give better economic and environmental policy guidance than the controversial estimates of the so-called external costs [48]? (2) How does the energy price influence the use of the different techniques of rational energy use, assuming cost-minimizing behavior of economic actors? To answer these questions, the main sources of physical and energy factual information, fundamentals of technical thermodynamics and the development of thermoeconomic optimization models were the works of Fricke and Borst [49], Baehr [50] and Foulds [51]. An introduction to energy/exergy analysis is given in [52].[8]

2.3.2 Pollutant Retention and Disposal

Pollutant Retention and Disposal (PRD) the burden on the biosphere from pollutants formed in energy conversion processes to levels below limits that society must set by consensus under risk assessments. Risk assessment is the most difficult part of the problem.

If the energy required for PRD is taken from fossil and nuclear energy sources, heat emissions are released into the environment due to the Second Law of Thermodynamics. As the Earth becomes more industrialized, they contribute to the approach of the heat barrier described in Sect. 1.1.2. The use of solar energy, which reduces the proportion of solar energy reflected by the Earth's system compared with the solar energy irradiated (albedo), also results in additional heat inputs.

The pollutant heat equivalent (PHE) is defined as the waste heat generated by the retention and disposal of a pollutant in a production process of given

[8]One of the pioneers of exergy analysis, van Gool, felt so strongly about the importance of energy, entropy, and exergy to the modern world that he completed his last work on them on his sickbed shortly before his death [53].

yield divided by the primary energy consumption of a production process of the same yield without pollutant retention and disposal.

In this sense, PHE is a thermodynamic environmental impact indicator that could be used as a guide for environmental impact taxes.

The heat emissions of a production process without PRD are usually equated to the enthalpy (calorific value) of the primary energy consumed. The energy analysis [52] is used to determine this. To calculate the PHE for known or projected processes of PRD, a method has been developed and exemplarily applied, that of energy analysis [54, 55]. For this purpose, one has to define the physical and chemical process boundaries, decide which indirect energy inputs and heat emissions should be considered, and establish rules for the generation of the energy and waste heat balances. A distinction must be made between substances that are intrinsic to the system and those that are not.

Systemic substances are present in the "biosphere" system as a permanent, passive substrate. If they occur in PRD processes, their consideration in the energy and material balances begins and ends with the state they assume under standard environmental conditions. For example, water vapor released to the environment during a PRD process is considered in the energy balance with the waste heat given by the condensation energy of the water vapor to water at ambient temperature. Non-system pollutants, on the other hand, must in principle be tracked along their entire path through the biosphere, beginning with degradation from natural deposits or production in chemical and nuclear reactions and ending with safe landfilling or conversion to native substances. For practical calculation, the path ends where the amount of the non-system substance falls below the valid limits.

The PRD ends with the landfilling of the pollutants. The landfills considered in the analysis are: (1) open landfills for inert and ecologically benign substances in non-volatile compounds that are stable for long periods under environmental conditions, e.g., calcium sulfate from desulfurization; (2) closed landfills for inert but potentially ecologically harmful substances that are stable for long periods under landfill conditions when placed in non-volatile forms, e.g., heavy metal; (3) off-earth space for aggressive, volatile and environmentally hazardous substances unless society tolerates their underground storage, e.g., spent fuel rods from nuclear power plants.

The PHE are calculated according to the following rules, which are only roughly outlined here.

1. Mechanical and electrical energy, used to overcome friction, becomes heat in its entirety. It is also counted as waste heat if it is first converted into potential energy that this does not do any useful work, but dissipates more or less quickly. The same applies to energy used for material transport within the biosphere. The total waste heat associated with the use of mechanical and electrical energy also includes all waste heat generated during the conversion of primary energy into the mechanical and electrical exergies according to known efficiencies.

2. Endothermic chemical processes emit the part of the process heat that is not converted into chemical binding energy as waste heat. In exothermic chemical processes, the enthalpy of reaction provides the waste heat.
3. The reduction in the efficiency of a production process as a result of PRD should be compensated for as directly as possible by increased primary energy input Q. The resulting additional waste heat ΔQ of the production process is included in PHE. It is determined by the primary energy (or process heat) required without PRD, Q_0, the efficiency without PRD, η_0, and the efficiency reduced by PRD, η, if the yield of the production process is to remain unchanged, so that $Q_0\eta_0 = Q\eta$. This gives $\Delta Q = Q(1 - \eta) - Q_0(1 - \eta_0) = Q_0(\eta_0 - \eta)/\eta$. If the compensation of a reduced efficiency is not possible by increased process heat input, then the total waste heat of a production process additionally required to compensate for the reduced productivity, which is also pollutant removed, must be taken into account.

The economic significance of limit values and PRD is demonstrated by the scandal surrounding German diesel cars. In the diesel engine, the air in the piston is heated by high compression to temperatures at which the injected diesel fuel ignites itself. The diesel engine is more energy-efficient, and therefore more suitable for reducing CO_2 emissions in traffic, than the gasoline-burning gasoline engine, for which the fuel is ignited by the spark plug after compression. However, the nitrogen oxide emissions[9] of modern German diesel engines in high-performance passenger cars can only meet the current statutory limits for NO_x if sufficient quantities of urea are injected into a special catalytic converter that converts the toxic nitrogen oxide into the system's own substances nitrogen and water. In 2007, the major German car manufacturers installed urea tanks with volumes of 17–35 L, which allowed ranges of between 16,000 and 30,000 km without refilling with urea. Then, in order to save space and costs, 8 L tanks were agreed. But this meant that, for the same range requirements, it was only possible to inject less urea than was necessary to comply with the NO_x limit values. This was achieved through illegal urea deactivation programs in the engine software. From 2015 onwards, these manipulations were successively uncovered and sanctioned. The penalties, compensation payments, image damage and sales losses in one of the most important German industries demonstrate the economic consequences of inappropriate handling of thermodynamic restrictions in the face of legal limits.

PRD for Coal-Fired Power Plant

As an example for the calculation of PHE, we consider a large hard coal-fired power plant fired with low-sulphur hard coal. Per standard cubic metre, its flue gas contains about 2000 mg SO_2 and 1000 mg NO_x (calculated as NO_2). According to the German Ordinance on Large Combustion Plants referred to in Sect 1.1.2 for power

[9]For NO_x production and effect (see Table 1.3). At high combustion temperatures, nitrogen in the air is also oxidized.

plants with a thermal capacity of more than 300 MW, the limit values per standard cubic metre are 200 mg NO_x and 400 mg SO_2. For CO_2 there are no legal limits yet. Following the recommendations of [12], a 66% reduction in emissions is assumed for the PHE of CO_2.

NO_x is removed from the flue gases by selective catalytic reduction with TiO_2 as catalyst and ammonia (NH_3) as reducing agent. The waste heat analysis results in a PHE of 23.71 $MJ/MWh_{thermal}$ = **0.66%** for 80% denitrification. Lime scrubbing process is widely used in Germany for removal of **SO_2** from power plant flue gases. For 95% desulfurization, waste heat analysis yields a PHE of 158.4 MJ/ $MWh_{thermal}$ = **4.4%**. **CO_2** Capture and Storage (CCS) was first investigated for three processes downstream of combustion, namely chemical scrubbing with alkanolamine [56, 57], freezing under pressure [57, 58] and physical absorption of CO_2 by Selexol [59]. Disposal options discussed included dumping the CO_2 in the deep sea or in empty oil and gas fields, and carbon recycling using solar carbon technologies and biomass-to-fuel conversion [60]. Alkanolamine scrubbing with 90% CO_2 capture would reduce power plant efficiency from 38% to 29%. If pressurized freezing and (ecologically risky) deep sea deposition are to reduce CO_2 emissions by 66% with undiminished electricity production, power plant efficiency would be reduced from 38% to 28%, and the PHE of CO_2 is 1.38 $GJ/MWh_{thermisch}$ = **38.8%**.

Between the years 2006 and 2014, Vattenfall Europe operated a pilot plant for CO_2 capture using the oxyfuel process, in which coal combustion takes place under pure oxygen, at the "Schwarze Pumpe" plant site in Lusatia. The plan was to permanently deposit the captured CO_2 in geological repositories such as empty gas fields or saline aquifers. The decision to shut down and completely dismantle the pilot plant on 1 September 2014 was justified by Vattenfall on the grounds of the political framework conditions in Germany. The know-how gained is now to be used by a Canadian company [61].

PRD for Nuclear Power Plant

Underground disposal sites have not yet been accepted as permanent storage sites for spent nuclear fuel rods from nuclear power plants for a number of reasons. The difficulties concern not the quantity but the risk assessment. A 1300 $MW_{electric}$ boiling water reactor consumes about 12,840 fuel rods annually during 7000 h of operation, producing 56 t of highly radioactive waste. A coal-fired power plant of the same annual production of electrical energy emits 8.5 10^6 t of CO2 annually. This CO_2 mass is more than 100,000 times the mass of the highly radioactive waste.

Pollutant heat equivalents of spent nuclear fuel rods have been calculated for their disposal into interstellar space [54, 55]. However, whether society would be prepared to bear the associated costs and risks is more than questionable. On the other hand, despite the phase-out of nuclear fission technology, Germany also has to take care of the permanent disposal of the approximately 30,000 m^3[10] of highly radioactive waste from the use of nuclear energy since its beginning in the 1970s until its end in 2022. If salt domes such as Gorleben are rejected as a final repository after careful exploration, the final disposal of radioactive waste in rock is an option that is now being considered. This would involve enclosing the radioactive waste in suitable containers, sinking them into holes drilled deep into the rock, and filling any remaining cavities with water-impermeable material. No nuclear-using country has yet pursued this option or estimated its energy requirements. But space disposal is likely to be the most energetically costly process, so it is on the thermodynamically safe side for the PHE calculation.

We consider rockets and electromagnetic material launchers for accelerating safely packed spent fuel to the minimum speed of about 11 km/s (escape velocity) required to leave the Earth's gravitational field.

For hydrogen-powered rockets, each carrying 350 kg of nuclear fuel rods into deep space, the PHE of highly radioactive nuclear waste would be about **0.27%**. But it would take 25,000 rocket launches a year to dispose of the spent nuclear fuel rods that all the world's nuclear power plants produced in the 1990s. That is unlikely to be feasible. In contrast, Henry Kolm, a physics professor at the Massachusetts Institute of Technology, had proposed using electromagnetic accelerators to launch the packaged radioactive waste directly from the Earth's surface into space [62]. Mass drivers operate on the principle of the electromagnetic linear motors that drive the maglev trains developed and tested in Germany and Japan and operated commercially in China. Prototypes of these mass drivers had been built by *Henry Kolm, Gerard K. O'Neill* (professor of physics at Princeton), and collaborators at Princeton University for the space industrialization [63–66] outlined in Sect. 6.2. The mass driver for nuclear waste disposal would draw its energy from huge capacitor banks charged by a 1000 $MW_{electric}$ nuclear power plant. The technical data are shown in Table 2.3.

Despite the high air friction, according to NASA studies, a steel cylinder equipped with a silicon carbide heat shield would not burn up during the passage through the dense Earth atmosphere, which would take only about 1.5 s. The steel cylinder would be able to carry a maximum of 40 fuel rods with a total mass of 175 kg. A maximum of 40 fuel rods with a combined mass of 175 kg could be transported in the loaded steel cylinder projectile with a total mass of 1000 kg, if a critical load of $6\ 10^8$ N/m^2 is not to be exceeded in the steel jacket at a 1000-fold gravitational acceleration. Projectile manufacture, production of the energy required for acceleration in the nuclear power plant with an efficiency of 33%, losses during acceleration and air friction would generate 233 GJ of waste heat per launch. For a 1300 $MW_{electric}$ boiling water reactor with 56 t of spent fuel rods produced per year,

[10] By 31 December 2010, 13,471 tonnes of irradiated fuel assemblies had been accumulated. The total volume of heat-generating radioactive waste from nuclear energy use to be disposed of is estimated at 29,030 m^3 in the forecast of the Federal Office for Radiation Protection after entry into force of the amended Atomic Energy Act of 6 August 2011.

Table 2.3 Technical data for the disposal of spent nuclear fuel rods into interstellar space by means of an electromagnetic mass driver with a thousandfold acceleration due to gravity (g) [62]

Projectile	Steel cylinder
Mass	1000 kg
Start speed	12.3 km/s
Upper atmospheric velocity (Escape velocity)	11 km/s
Kinetic energy at start	76 GJ
Melting loss of the SiC shield	3% of the mass
Energy loss	20%
Acceleration	1000 g
Length of the material slingermass driver	7.8 km
Duration of acceleration	1.2 s
Mean force	$9.81\ 10^6$ N
Average power	60 GW
Charging time of the capacitors	1.5 min

321 firings/year would be required and about 75 TJ of waste heat would be released per year as a result. In relation to the approximately 27 TWh of thermal energy generated per year from nuclear fission, this gives a PHE of **0.08%**. If one wants to minimise the risks in the event of a mishap including a crash of the projectile and instead of 40 fuel rods only one is packed into the projectile, the PHE of space disposal of spent nuclear fuel rods increases to **3%**.

If one wanted to dispose of the spent fuel rods of 160 nuclear power plants, each of which has a capacity of 1300 $MW_{electric}$, so that their total capacity corresponds to 56% of the global nuclear generation capacity of 371.7 $GW_{electric}$ existing in 2007, one would need more than 50,000 electromagnetic launches per year, each with a payload of 40 fuel rods. If a mass driver were used every 12 min day and night [62], this would amount to 43,800 launches per year. Each launch produces an explosive shock wave in the atmosphere with an energy content ten times greater than the energy of 0.1 MWh in the main branch of an average lightning discharge, is likely to disrupt the ionic balance and chemistry of the atmosphere near the launch site. The consequences are unknown. But to be on the safe side, one should also include in PHE the relatively small amount of energy required to transport all system components to desert launch sites.

The pollutant heat equivalents of SO_2-, NO_x- and CO_2-disposal from coal-fired power plants are at least a factor of 10 greater than those of the space disposal of highly radioactive waste from nuclear power plants. Thus, they give an impression of the correspondingly different degree of approach to the heat wall in the PRD of coal-fired and nuclear power plants. An environmental tax burden on energy sources oriented to the PHE would take this into account.

However, the PHE method does not solve the problem of risk assessment and evaluation of all other side effects of fossil and nuclear energy use. The following incident shows how far risk assessment can go in the case of nuclear energy. Once, when the conversation among friends turned to the question: "Where to put the nuclear waste?", the study on the disposal in space was also mentioned. Someone with a degree in natural sciences was outraged: "First you physicists pollute the earth with atomic bomb explosions and nuclear power stations, and now you want to contaminate space with radioactivity!"

2.3.3 Rational Use of Energy

If all devices for converting solar radiation into electrical energy, space and process heat were one day manufactured from solar energy, life cycle emissions such as those given in Table 1.2 for CO_2 should no longer be a problem. But there is still a long way to go. Moreover, Table 2.4 shows CO_2 emissions per capita and year according to both the *consumption principle* and the *production principle*. The former attributes the emissions associated with the production of imported goods to the importer, the latter to the exporter. Official emission statistics are based on the production principle. In them, wealthy Central European countries in particular, which relocate energy-intensive industries abroad and import their products, come off too well.

Also, ecological damage from biomass that does not come from sustainable forestry, as well as future environmental damage from the electronic waste of photovoltaics, is not given enough credit in discussions of sustainable development. Be that as it may, the use of techniques that reduce primary energy demand while providing an undiminished energy service is one of the best options for reducing emissions in many respects. The question is what are the associated costs?

The results of thermo-economic optimization studies depend essentially on the structure, the limits, and the energy demand profile of the system under consideration. In addition, when considering local systems, the energy supply structure of the superordinate system, e.g. the country in which they are embedded, must be taken into account. Tables 2.5, 2.6, and 2.7 provide information on the structure of Germany's energy supply between 1990 and 2016.

The two optimization studies on rational energy use (REU), the results of which are discussed here, are based on the energy supply structures of the Federal Republic

Table 2.4 CO_2 emissions by attribution principles, in tonnes of CO_2/person/year (2011) [67]

Country	Consumption principle	Production principle
Kuwait	34.5	33.9
Australia	29.4	33.4
USA	27.9	23.5
Switzerland	23.0	7.3
Austria	21.5	12.8
Germany	18.3	13.2
Russia	14.0	19.4
Iran	11.0	10.6
PRC	8.4	9.5
India	3.3	3.5

Table 2.5 German gross electricity generation the years 1990, 2000, 2010 and 2016, in terawatt hours (TWh) [68]

Energy source	1990	2000	2010	2016
Lignite	170.9	148.3	145.9	150.0
Nuclear energy	152.5	169.6	140.6	84.6
Hard coal	140.8	143.1	117.0	111.5
Natural gas	35.9	49.2	89.3	80.5
Petroleum products	10.8	5.9	8.7	5.9
Wind energy onshore	n.a.	9.5	37.8	65.0
Wind energy offshore				12.4
Hydropower	19.7	24.9	21.0	21.0
Biomass	n.a.	1.6	28.9	45.6
Photovoltaics	n.a.	0.0	11.7	38.2
Household waste	n.a.	1.8	4.7	6.0
Other energy sources	19.3	22.6	26.8	27.5
Total	549.9	576.6	632.4	648.4

Table 2.6 Percentage shares of energy sources in Germany's gross electricity generation in 1990, 2000, 2010 and 2016 [68]

Energy source	1990	2000	2010	2016
Lignite	31.1	25.7	23.1	23.1
Nuclear energy	27.7	29.5	22.2	13.1
Hard coal	25.6	24.8	18.5	17.2
Natural gas	6.5	8.5	14.1	12.4
Petroleum products	2.0	1.0	1.4	0.9
Wind energy onshore	n.a.	1.6	6.0	10.2
Wind energy offshore				1.9
Hydropower	3.6	4.3	3.3	3.2
Biomass	n.a.	0.3	4.6	6.9
Photovoltaics	n.a.	0.0	1.9	5.9
Household waste	n.a.	0.3	0.7	0.9
Other energy sources	3.5	3.9	4.2	4.3
Total	100.0	100.0	100.0	100.0

Table 2.7 Primary energy consumption in the Federal Republic of Germany [69]

Energy source	1990	2000	2010	2016	2016 (%)
Mineral oil	5217	5499	4684	4563	34.0
Natural gas, -oil	2293	2985	3171	3043	22.7
Hard coal	2306	2021	1714	1635	12.2
Lignite	3201	1550	1512	1525	11.4
Nuclear energy	1668	1851	1533	927	6.9
FRE	139	290	1160	1163	8.7
PWH	58	127	254	529	3.9
FTBE	3	11	64	200	1.5
Other	22	68	254	242	1.8
Total	14,905	14,402	14,217	13,426	100.0

Figures in petajoules (PJ); *FRE* fuels from renewable energy sources, *PWH* photovoltaics, wind, hydropower, *FTBE* foreign trade balance electricity

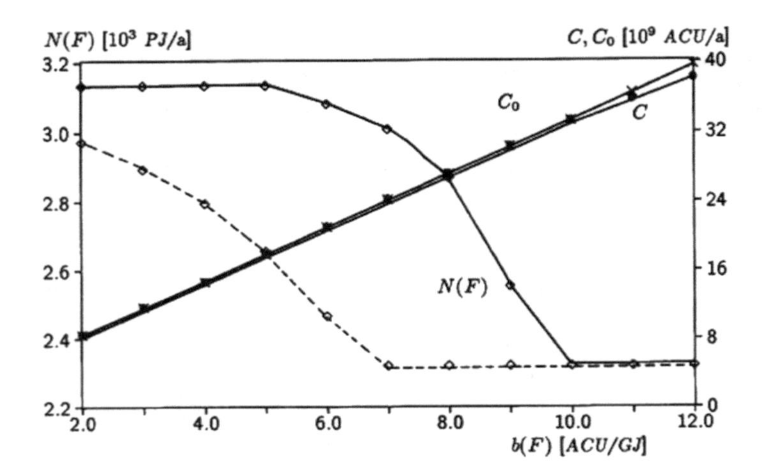

Fig. 2.2 Annual primary energy demand $N(F)$ of the (old) FRG and the annual costs C associated with its coverage *with* and C_0 *without* heat recovery. Dependence on the fuel price $b(F)$ according to the static optimisation model *ecco* [70]. The solid (dashed) $N(F)$ curve applies at the cost ceiling $C \leq C_0$ ($C \leq 1.1\ C_0$). The minimum value of $N(F)$, above which the cost ceilings no longer have an effect because the energy savings potential is exhausted, is 2321 PJ/a. A primary energy price of $b(F) = 4$ ACU/GJ is roughly equivalent to the 1986 oil price of US$24 per barrel. As Fig. 3.1 shows, this price is close to the 2014 oil price in inflation-adjusted US$

of Germany (FRG) in the 1980s and 1990s. The aim of both studies was to calculate, under optimistic assumptions, the maximum possible reductions in primary energy use for given energy demand profiles.

The static model of energy, cost and CO_2 optimization *ecco* was used to determine the thermodynamic limits of energy saving in the old FRG and the associated costs depending on the energy price; Fig. 2.2 shows one result. From *ecco*, the model *deeco* of dynamic energy, emission and cost optimization was developed and applied to a model city "Würzburg". This model city would be built "on a greenfield site". Its fluctuating demand for heat and electrical energy is based on the demand of the real existing city of Würzburg as determined by the Würzburg public utility company for the year 1993. Figures 2.3 and 2.4 show some results.

Both studies exemplify the influence of energy prices on energy, emission and cost optimisation as well as the synergy and competition effects of REU technologies for model systems whose energy demand is oriented towards that of real systems.

The underlying thermodynamic and economic principles should be taken into account in the conversion and expansion of modern energy supply systems.

The estimation of the thermodynamic limits of REU through heat recovery by means of the *ecco* model is based on the demand structures for industrial process heat with temperatures between 50 °C and 1700 °C documented by institutes such as the Munich Research Centre for Energy Economics. These were converted to

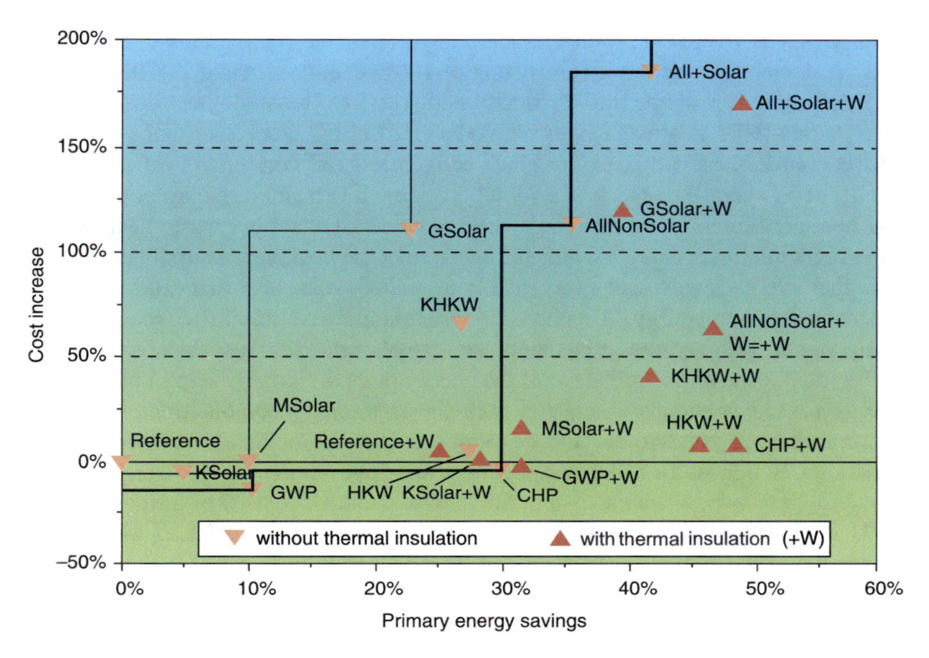

Fig. 2.3 Relative reduction of primary energy use and the associated cost increases in "Würzburg" according to the *deeco* model of dynamic energy, emission and cost optimization [75]

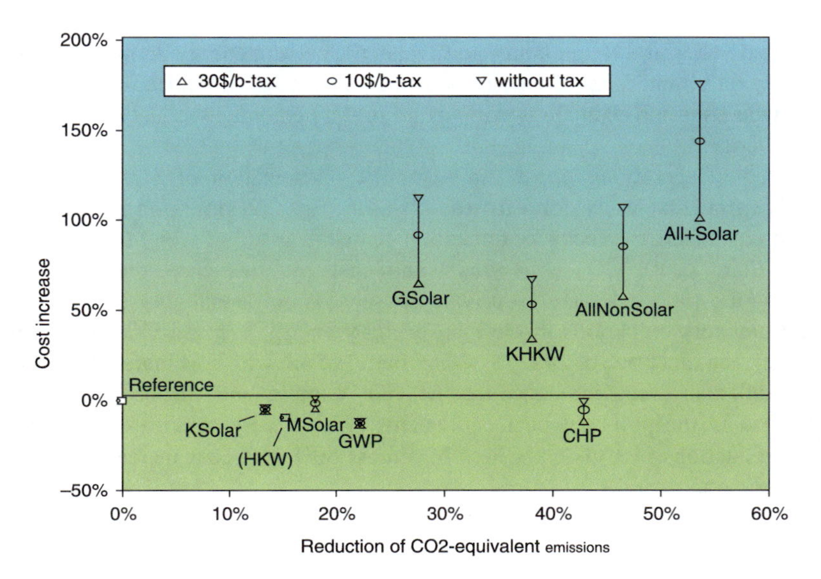

Fig. 2.4 Relative reduction in "Würzburg" CO_2 equivalent emissions and associated cost increases for REU scenarios and combined CO_2 energy taxes in US$ per barrel of oil according to the *deeco* model [75]

enthalpy-exergy demand profiles for Germany, Japan, the Netherlands and the USA, and the demand for electricity consisting of 100% exergy was added [70, 71].

For simplicity, a single fuel (*F*) is assumed. The REU technologies considered are heat pumps (HP), heat exchanger networks (HE) of an average distance of 50 km between waste heat producers and heat consumers, and combined heat and power (CHP) of an average distance of 25 km between waste heat producing power plants and heat consumers. Heat exchange networks and CHP supply waste heat directly from higher to lower exergy levels. Indirectly, electric heat pumps supply consumers at higher exergy levels with exergetically upgraded waste heat that originates from sources of lower exergy. The objective function to be minimized contains the components "primary energy demand" and "total costs" of the system.

Figure 2.2 shows the influence of the energy price $b(F)$ on the exploitation of the energy saving potentials associated with the technology combinations, assuming cost-minimizing behavior of the economic actors. Here, in a first scenario, the constraint is imposed on the optimization that investments in REU technologies may only be made to such an extent that the total annual costs C of the system *with* REU do not exceed the total costs C_0 of the system *without* REU. The cost is composed of the annual investment cost (using an annuity factor of 0.175), maintenance cost and energy cost. In this case, the full savings potential of REV techniques is only realized at a fuel price of 10 ACU/GJ ($\approx$60$\$_{1986}$ per barrel of oil). If, on the other hand, a cost cap of 1.1 C_0 is allowed in a second scenario, e.g. if REU techniques are supported by government subsidies, the energy demand minimum of 2321 PJ/a is already reached at a fuel price of 6.5 ACU/GJ. (This is 74% of the primary energy demand without REU. For the enthalpy-exergy demand profiles of the United States, the Netherlands, and Japan, the percent energy demand minimums are 70%, 66%, and 53% of energy demand without REU, respectively [73]). The percentage emission reductions correspond to the percentage reductions in primary energy demand.

Instead of specifying upper limits for the optimization of a multicomponent objective function, as is done for the costs in Fig. 2.2, emission reductions can also be calculated by vector optimization with linear target weighting of primary energy (hard coal), CO_2 emissions[11] and costs. In this case, the *ecco* model, including the energy demand of private households in the enthalpy-exergy demand profile, delivers reductions in German primary energy use and CO_2 emissions of 41% each at a fuel price of 24 US$\$_{1986}$ per barrel of oil, with cost increases of 44%. If the model also takes into account the CO_2 capture and disposal technologies mentioned in the pollutant heat equivalents (with a very rough estimate of their costs), reductions in CO_2 emissions of almost 80% and cost increases of around 100% are obtained [72].

Even the result of the relatively simple optimization model *ecco* shown in Fig. 2.2 raises the question that energy and environmental policy always faces in a real market economy: **How should incentives to reduce emissions be provided—by**

[11] CO_2 emissions of 0.31 kg/kWh^{thermisch} are assumed for the combustion of hard coal without REU.

increasing the price of energy through energy taxes or by subsidizing the use of emission-reducing technologies? One often gets the impression that voters and politicians prefer subsidies to noticeable energy tax increases. Market-oriented economists, on the other hand, consider subsidies to be questionable because they may rely on the wrong technology and push the system onto a suboptimal development path. And economists are almost unanimous in their view that the third alternative, namely order and bans (e.g. on energy sources and energy technologies), can only be justified in the case of imminent danger.

Whereas the *ecco* model uses the static enthalpy-exergy demand profiles of entire economies to provide the costs of reducing emissions through REU, taking a highly simplified approach to the limits of what is thermodynamically possible, the much more complex *deeco* model allows the costs of reducing emissions through REU to be calculated for regional energy systems, taking into account the temporal fluctuations in the demand for heat and electricity as well as the outdoor temperature and solar radiation. Renewable energy sources also come into play, and differentiation is made between fossil energy sources. Reliable data could be collected for the city of Würzburg. The model city "Würzburg" optimized with its data will be referred to simply as the city in the following. In it, heat and electrical energy are mainly demanded by households and small consumers. The core area is the inner city, which is supplied with heat by a combined heat and power plant via a district heating network. The peripheral area is the rest of the urban area. The optimisation is based on measured time series (a) of the heat demand per unit time: (1) of the core area on average 43 MW, peak 125 MW, (2) of the peripheral area on average 129 MW, peak 374 MW, (b) of the electrical energy demand of the city (on average 69 MW, hourly peak 122 MW), (c) of the solar irradiation (annual average 1160 kWh/(m^2a)), and (d) of the outdoor temperature (annual average 9 °C).

The city's energy system to be optimised generates all its heat from coal, oil and gas. Electrical energy is partly generated by the city, partly drawn from the German power grid or fed into it. In the reference scenario, the city is supplied with heat from oil boilers and electrical energy from the German power grid. The combination of techniques to be selected by the optimization program *deeco* for the dynamic minimization of the objective function with the components non-regenerative primary energy, costs, as well as CO_2-, NO_x- and SO_2-emissions form scenarios. In this context, *deeco* provides special versions of the basic technologies already used by *ecco* (HP, HE, CHP) of REU as well as their combinations—also with direct heating by oil and gas boilers. In addition, there are solar thermal systems and thermal insulation *W*. The latter reduces the space heating requirement by 50%. Around 10 of the 22 scenarios calculated are outlined in Table 2.8.

Figure 2.3 shows for the scenarios considered the primary energy savings achievable with the technologies used and the associated cost increases as a percentage of the primary energy demand and costs of the reference scenario. The thick "trade-off" staircase curve indicates the minimum cost increase at which a given primary energy saving can be achieved without thermal insulation. The results for the individual scenarios without and with thermal insulation (+W) are scattered around the thick trade-off curve. The corresponding trade-off curve for the scenarios with thermal

Table 2.8 REU scenarios for "Würzburg"

Scenario	Techniques
Reference	Oil boiler
KHKW	Gas-fired small heating plants, gas boilers
CHP	Local heating from gas-fired combined heat and power plants, gas boiler
HKW	District heating from hard coal-fired cogeneration plants and from peak gas boilers
GWP	Local heating from outdoor air gas heat pumps and gas boilers
KSolar	In the peripheral area 258 small solar local heating systems without heat storage (contribution margin 7%), local heating from gas condensing boilers; in the core area oil boilers
MSolar	In the peripheral area 258 medium solar local heating systems with heat storage (contribution margin 25%), local heating from gas condensing boilers; in the core area oil boilers
GSolar	In the peripheral area 258 large solar local heating systems with seasonal heat storage (contribution margin 71%), local heat from gas condensing boilers; in the core area oil boilers
AlleNichtSolar	All non-solar technologies plus electric heat pumps
Alle+Solar	All as in AllNonSolar plus large solar local heating systems as in GSolar

In all scenarios, electricity procurement from the German grid is permitted

insulation, which is not plotted, lies in the range of significantly higher energy savings.

In the CHP scenario, it is possible to save 30% of the primary energy used in the reference case without increasing costs. CHP plants cover 57% of the heat demand and 86% of the electricity demand. The remaining 43% of the heat demand is supplied by gas boilers. The HKW scenario is only slightly less favourable. On the other hand, in the AllNonSolar scenario, when all non-solar technologies can be chosen from, KHKW and CHP plants supply 73% of the heat and 86% of the electricity demand. Gas condensing boilers almost completely satisfy the remaining heat demand with a share of 21%. All other technologies are only rarely used during the year, but hit with their full investment and maintenance costs, so that a cost increase of 114% compared to the reference scenario only results in an increase in primary energy savings of 6% points compared to the CHP scenario. This is a typical example of how many REU technologies compete with each other when they interact, so that their savings potentials cannot be fully realized. For the same reason, the interaction of large solar collectors and seasonal thermal storage with all non-solar technologies in the All+Solar scenario leads to increases in primary energy savings to only 42%, while costs skyrocket to 186%.

According to the thinly drawn trade-off staircase curve for the solar scenarios, the MSolar scenario delivers a 10% primary energy saving without any cost increase, while in the GSolar scenario, due to heat losses in the collectors and the seasonal storage, the primary energy saving is only 23%, but the costs increase by 111%.

Figure 2.4 shows the relationship between emission reduction and energy system costs. The results are based on the minimization of CO_2-equivalent emissions, which are obtained by summing up all climate-relevant emissions with relevant weight factors. (Thermal insulation measures not considered here have the same shifting effects as in Fig. 2.3). The cost calculations are based on energy prices in the mid-1990s. In addition, two further energy tax rates are assumed: a combined CO_2/energy tax of US$10 per barrel of crude oil in accordance with the EU Commission's proposal for a directive [76] and a tripling of this tax in line with the IEA CO_2 tax scenario [77]. The specified tax rates are related in a ratio of 1:1 to the calorific value and the specific CO_2 emissions of the crude oil. This results in energy and CO_2 tax rates that are used to tax the other energy sources according to their energy and carbon content [78, 79]. The relative cost increases at different tax rates shown in Fig. 2.4 show that, on the one hand, the distance of all significantly emission-reducing scenarios to the economic viability threshold is significantly reduced by the assumed energy price increases, and on the other hand, a breakthrough to economic viability is only achieved in the MSolar scenario (The CHP, GWP and KSolar scenarios are more cost-effective than the reference scenario anyway).

If one wanted to implement the results of the *deeco-based* optimization, one would have to build a new city. In this respect, the results (only) show what would have been possible in the 1990s, at best, in terms of emission reductions at the municipal level, and at what investment and energy costs. Later studies with *deeco* on energy, emission and cost optimization in new buildings [80] and conversions [81] of residential complexes confirm three central findings of the studies discussed here: (1) The maximum energy savings potentials of individual REU techniques do not add up at all when these techniques are combined. On the contrary, they often compete with each other, e.g. for heat supply, so that the most cost-effective combination of technologies comes into play the most, and the other technologies are hardly, or not at all, used, while their investment costs have to be paid in full. (2) Thermal insulation of buildings allows significant energy savings with moderate cost increases. (3) As energy prices rise, the energy costs saved by REU increasingly offset the investment costs of REU techniques. Thus, the use of emission-reducing REU techniques in economies that seek cost minimization is more favored by higher energy prices than by low ones.

3 Wealth Creation and Growth

> The ideas of national economists – whether right or wrong – are far more influential than is generally believed. In fact, the world is hardly governed by anything else. Practitioners who believe themselves utterly free from all intellectual influence are usually but slaves to some late national economist. Mad politicians who hear voices in the air usually get their nonsense from some academic hack of yesteryear. I'm sure the influence of vested rights and interests is vastly overstated compared to this slowly but steadily growing influence of ideas. Such things don't happen instantly, of course; such a process takes time. In the field of economic and political philosophy, there are not many who are influenced by new theories after they are older than 25 or 30. Therefore, it is not very likely that officials, politicians and even agitators apply the latest ideas to current events. But sooner or later it is the ideas and not the various interests that are dangerous – whether for good or for evil. (John Maynard Keynes)[1]

3.1 Economics: Orthodox

"Economics …: the oldest of the arts, the youngest of the sciences – perhaps even the queen of the social sciences". This is how the Nobel Prize winner in economics, Paul A. Samuelson, describes his science in the introduction to Volume I of his text book *Volkswirtschaftslehre* [33].

Adam Smith's 1776 work *The Wealth of Nations*, already mentioned in Sects. 1.2.3 and 2.3, established classical national economics. Since then, the agrarian economic system has been transformed into an industrial one. The classical production factors of land and labour are no longer sufficient to understand the processes of value creation. Therefore, the neoclassical theory, which dominates especially in quantitative analyses of economic growth and is based on the ideas of classical national economics, has placed "technical progress" alongside the factors of capital

[1]These last lines from John Maynard Keynes' 1936 book The General Theory Of Employment, Interest and Money are taken from [33, Vol. I, p. 32]

R. Kümmel et al., *Energy, entropy, creativity*,
https://doi.org/10.1007/978-3-662-65778-2_3

and labour.[2] Similarly, the new, so-called "endogenous" growth theory emphasises "knowledge".

However, since "technical progress" is not specified causally and "knowledge" is not quantified, a veil remains over the driving forces and inhibiting limits of economic growth. In the tradition of Adam Smith, textbook economics today is still primarily interested in the behaviour of economic actors in competition and the formation of prices in markets. The physical sphere of production of goods and services lies on the margins, if not outside the field of vision of most economists. If this physical sphere nevertheless comes to the fore, as in the case of the anthropogenic greenhouse effect, for example, there is a great danger that short-sighted assessments based on current market prices will lead to misjudgements of future developments with serious economic and social consequences.

Thus, many economic models evaluate the damage caused today but only occurring in the future according to what we call the "Esau principle". The individual time preference referred to by this has existed since biblical times. The Book of Genesis in the Old Testament portrays it in its narrative of Esau's sale of the birthright to Jacob for a dish of lentils. Esau had been born the son of Isaac and grandson of Abraham before his twin brother Jacob. He possessed the right of the firstborn, the preferential right to the inheritance. One day he came home from the hunt hungry, where Jacob was cooking a lentil puree. Esau said to Jacob, "Let me have a quick taste of that red food there, for I am exhausted." Jacob replied, "Sell me your birthright today." Esau, in response: "I am wandering about, and yet I will die! What good is the birthright to me?" And to satisfy his present hunger, he sold a promise for the future. Jacob received the paternal blessing intended for his brother and became the progenitor of Israel.

In accordance with the time preference of Homo sapiens, which has been practiced since time immemorial, present benefits are valued more highly than future benefits or even harm—despite the economic ethicists who speak of the scandal of discounting the future [82].

If, for example, global damage amounting to US$2000 billion were to occur in 150 years as a result of the anthropogenic greenhouse effect due to melting of the West Antarctic ice shelf and flooding of low-lying coastal areas,[3] which would correspond to about twice the gross domestic product (GDP) of the USA in 1971, and if this damage were discounted to the present using a discount rate of 4%, it would only correspond to just under six per mille of the GDP of the USA.[4] Investing more to compensate for or avert these future losses today would not be economically rational.

[2]Land, which can no longer be reproduced, plays virtually no role in questions of industrial economic growth.

[3]This estimate from the 1970s was made by climate researchers Chen and Schneider of the National Center for Atmospheric Research in Boulder, Colorado.

[4]At a discount rate of 7%, which was the interest rate for long-term investments before the central banks' zero interest rate policy after the first severe recession of the twenty-first century, the discounted losses would be a factor of 75 lower.

A second example of monetary myopia in the view of price and value is narrated by the former member of the Board of Directors of the World Bank, Herman Daly, in his article "When smart people make dumb mistakes", [83] in which he reports statements of the economists W. Nordhaus (Yale University), W. Beckerman (Oxford University) and T. C. Schelling (Harvard University, former President of the American Economic Association, Nobel Prize for Economics 2005). These highly respected and influential representatives of their field have assessed the risks of the anthropogenic greenhouse effect (AGE) from an economic perspective. They take as their starting point the fact that agriculture currently contributes only 3% to the Gross National Product (GNP) of the USA—its contribution is similarly low in the other industrially highly developed OECD countries—and they assume that agriculture is practically the only economic sector affected by the consequences of the AGE (Whether this assumption is justified remains to be seen). They thus conclude that even in the event of a drastic collapse in agricultural production, only insignificant welfare losses are to be expected: After all, even if agricultural production fell by 50%, GDP would fall by only 1.5%; if agricultural production were drastically reduced by climate change, the cost of living would rise by only 1–2%, at a time when per capita income would probably have doubled [In the original text, "there is no way to get a very large effect on the US economy" (Nordhaus), "even if net output of agriculture fell by 50% by the end of next century this is only a 1.5% cut in GNP" (Beckerman), and "If agricultural productivity were drastically reduced by climate change, the cost of living would rise by 1 or 2%, and at a time when per capita income will likely have doubled" (Schelling)]. This risk assessment misses the fact that if there is a drastic shortage of food, its prices will naturally explode and drive up agriculture's currently rather marginal contribution to GDP. It seems to have been forgotten that severe economic crises have always been accompanied by famines.

Textbook economics rates the importance of energy for the production of goods and services similarly low as food. In both cases, the current market price and the value for life and the economy are far apart.

There is nothing intrinsically wrong with that. On the contrary: after overcoming the consequences of the Second World War, a golden age dawned for the citizens of the Western industrialised countries. Never in the history of mankind had so many been so well off, because essential basic material needs could be satisfied at ever lower cost. Table 2.2 shows examples. Cheap energy led to decreasing costs of food production with increasing mechanization of agriculture and the extensive elimination of the human cost factor from agriculture. Therefore, the share of agriculture in the overall economic value added decreased. In 1950, five million people in the Federal Republic of Germany, or 25% of all employed persons, worked in agriculture and produced 11% of the gross domestic product (GDP). In the 1990s, by contrast, only about 3% of the labour force was employed in the agricultural sector, whose share of German GDP has fallen to about 1% (Tables 3.2 and 3.3).

It becomes problematic, however, if the current low prices of food and energy are deceptive about their tremendous value in terms of benefits to life and industrial production. Of course, mainstream economists will assure us that their science is not subject to such deception. Rather, they say, it distinguishes between marginal utility and total utility. Marginal utility is the utility of the last unit of an economic good that is demanded, while total utility is the total contribution of a good to economic well-being.[5] After all, they pointed out, neoclassicism had been resolving the value paradox of water and diamonds for a good 100 years, which since Adam Smith had been a famous problem and had already caused him grief in *The Wealth of Nations*. How could it be explained, he asked then, and many others after him, that water, although so useful that no life is possible without it, has such a low price, while diamonds, which are completely unnecessary, fetch such a high price?

To this, a textbook like Samuelson's answers today: diamonds are very *scarce,* and the cost of producing *additional* diamonds is high; water, on the other hand, is relatively *abundant*, and its cost is quite low in many zones of the earth. Moreover, the total utility of water does not determine its price or its demand. Only the relative marginal utility and the cost of the last unit of water determine its price. And why? Because people have the freedom to buy or not buy that last small unit of water. Therefore, if its price is above marginal utility, that last unit of quantity cannot be sold. And for this reason, the price must fall until it reaches exactly the level of marginal utility. In addition, each unit of water is exactly equal to all the others, and since there is only one price in a competitive market, each unit must be sold at exactly the price obtained by the last useful unit [33, Vol. II, pp. 88, 89].

The neoclassical insight that the last unit demanded of a good determines the price of all units of that good is considered so significant that it is sometimes called the "marginal revolution."

Marginal, i.e. very small changes are mathematically captured in the infinitesimal calculus by differentials, larger changes by integrals. *Newton*, alongside *Leibniz,* developed the infinitesimal calculus to describe the forces and motions of mechanical systems. Newtonian mechanics was so successful and fascinated so many scientists in the nineteenth century that its formalism was used as a model for the mathematical formulation of other scientific disciplines, including neoclassical economics. In neoclassical economics, the extremum principles used in physics play an important role. In Sect. 3.4, however, we will see that *technological constraints* not taken into account by textbook economics prevent the neoclassical optimum from being reached at current energy prices. The same follows from the optimization of overall welfare [2, 102, 124]. This will show how the adoption of *formal* aspects of *mechanics* by economics becomes compatible with the economically so important *contents* of *thermodynamics* when technological constraints are taken into account.

Ignoring technological constraints leads to a drastic underestimation of the importance of energy as a production factor in neoclassical textbook economics, as we will see in more detail in Sects. 3.3 and 3.4. This underestimation of energy has not only theoretical but also practical consequences. Firstly, as already indicated, it

[5] Why Beckermann, Nordhaus and Schelling did not make this distinction when assessing the consequences of climate change gives H. Daly reason to speculate about the effect of the "dogma" of the necessity of constant economic growth on the scientific rationality.

leads to the anthropogenic greenhouse effect being seen as only a minor problem in economic terms. For if energy were unimportant compared to the production factors labour and capital, the problem could be solved very simply by reducing energy use without significantly affecting the production of wealth. Second, the underestimation of energy as a factor leads to too little attention being paid to energy security in the energy policy debate, again with potentially serious consequences: After all, without energy, and electricity in particular, not only large parts of production but also machine-based communication processes and almost all public life come to a standstill. And thirdly, the economic consequences of energy or oil price shocks cannot be approximated with a low weighting of energy as a production factor. We will discuss this in more detail below.

While comparatively inert textbook economics still has a blind spot with regard to the importance of energy, the facts are increasingly being recognised by economic researchers. In his newsletter of 23 February 2018, for example, the chief economist of the Handelsblatt spoke of a "serious deficit in the standard models of economics" [84] with regard to the disregard or low weighting of energy as a production factor compared to labour and capital, and referred to a recent analysis for the Handelsblatt Research Institute [85].

3.2 Oil Price Shocks

Three recessions have shaken the world economy since the end of the Second World War. They were accompanied by the sharp fluctuations in the price of crude oil shown in Fig. 3.1 and can be seen in the examples of the USA and the Federal Republic of Germany (FRG) in Fig. 3.2. They have stimulated new thinking about economic growth because neoclassical growth theory, as part of the orthodoxy of economic science, has problems understanding the observed economic development.

Since the early 1950s, the world market price of a barrel (157 litres) of crude oil had fallen. It could be easily and cheaply extracted from the previously discovered vast oil fields in the Near and Middle East, Indonesia and the Americas. Together with the investments of the Marshall Plan, crude oil made an important contribution to the reconstruction of devastated Europe. Since then, as Fig. 2.1 shows, it has been the most important energy source in terms of volume. It was, and still is, indispensable for the transport of goods and thus for world trade.

The phase of steady economic growth of the post-war period, with annual growth rates of up to 7%, came to an abrupt end in the early 1970s. Worst of all was the first oil price shock in 1973–1975, triggered by the Yom Kippur War in October 1973. At that time, Israel had been unexpectedly attacked by Egypt and Syria on the Jewish

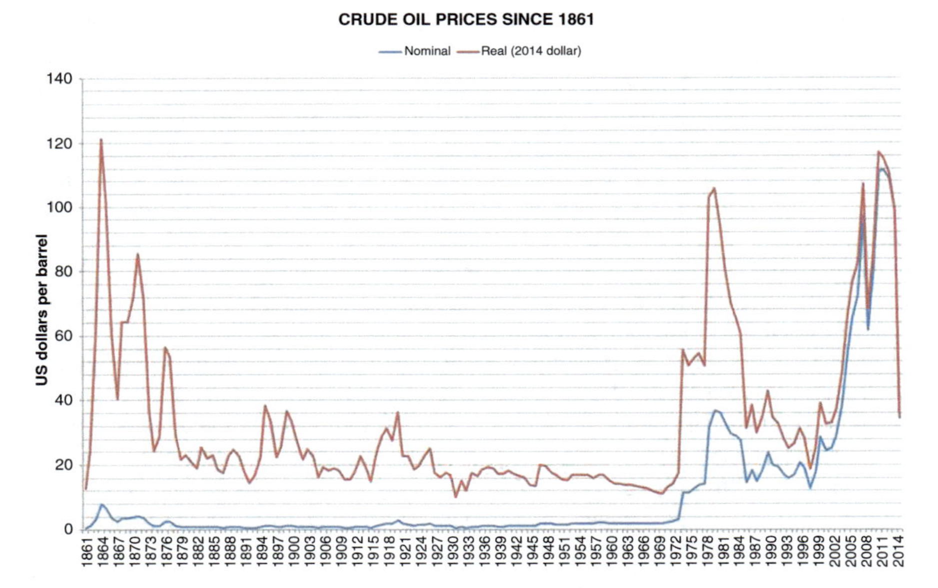

Fig. 3.1 Development of the average annual price of a barrel of crude oil between 1861 and 2014. Upper curve: adjusted for inflation in US_{2014}. Lower curve: in daily dollar prices. (Data from BP Workbook of Historical Data [86])

Day of Atonement, Yom Kippur. After initial great distress[6] it finally triumphed. With a boycott of deliveries, the oil-producing Arab states tried to force the non-socialist states of Europe, North America and Japan, which were considered friendly to Israel, to change their previous position in the Middle East conflict between Israel and its Arab neighbours, which had been smouldering since 1948: The Organization of Petroleum Exporting Countries (OPEC) drove up the price of oil on the world market from 16 $US\$_{2014}$ per barrel of crude oil in 1973 to 56 $US\$_{2014}$ in 1975, and economic activity in market economies collapsed worldwide (The "socialist" planned economies fared differently, as will be discussed below). The declines in energy use and economic growth between 1973 and 1975 in the Federal Republic of Germany and the USA are shown in Fig. 3.2; how Japan was affected is shown in Fig. 3.3. This is sometimes called the first energy crisis.

But "oil price shock" better points to the psychological-technical mechanism of the economic slump. For entrepreneurs feared that the fuel for their heat engines, which had been flowing so abundantly and cheaply until then, would run out, so that some of them would have to be shut down. So, in order to avoid overcapacities, they cut back their investments—in the industrial sector "goods producing industry" of the FRG from 90 billion DM in 1973 to less than 50 billion DM in 1975 [88, 89]. In

[6]Credible sources speak of the fact that after Israel's great initial losses it almost came to the use of nuclear weapons, which was only prevented by massive deliveries of war material on the part of the USA.

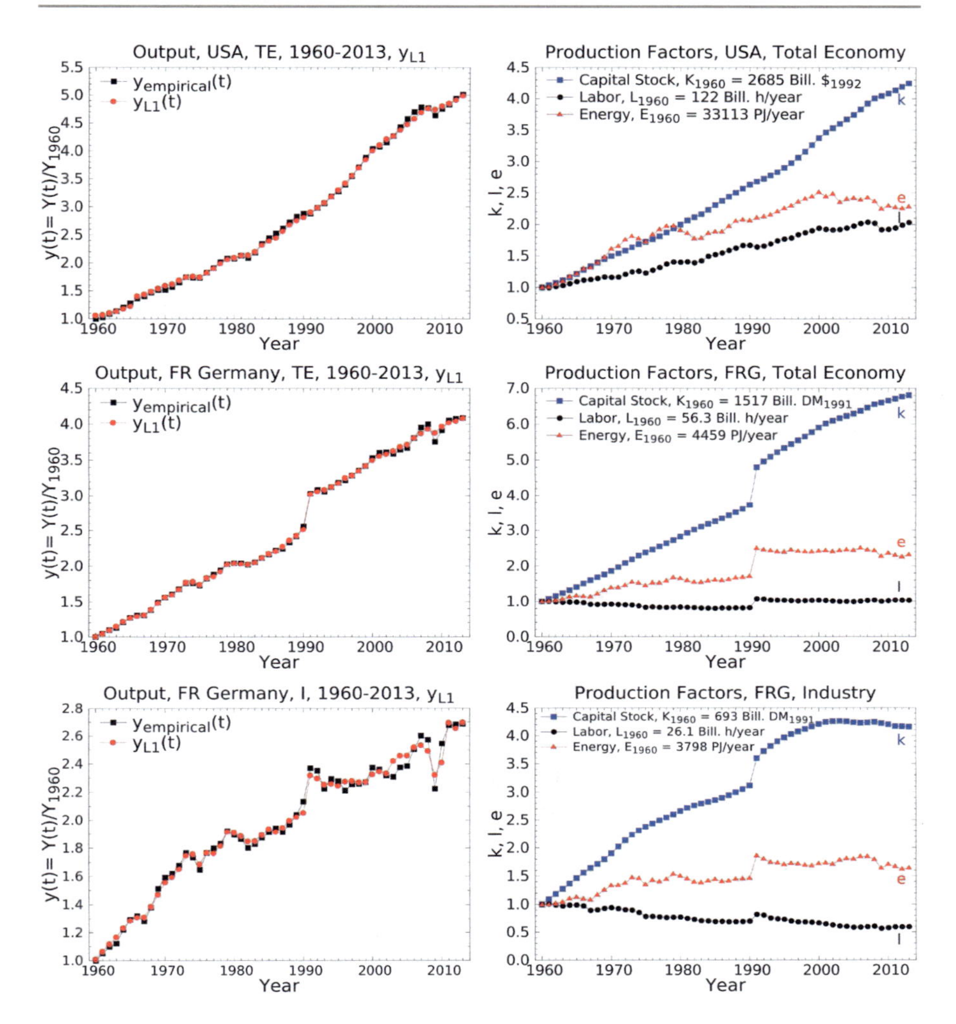

Fig. 3.2 Economic growth in the USA and the Federal Republic of Germany (FRG) between the years $t = 1960$ and $t = 2013$. Top: USA, total economy; middle: FRG total economy; bottom: FRG industry ("goods producing industry"). Left: (relative) value added (output) $y(t) = Y(t)/Y_{1960}$; empirical (squares) and theoretical (circles). Right: empirical production factors normalised to the base year 1960 capital stock $k = K(t)/K_{1960}$ (squares), labour $l = L(t)/L_{1960}$ (circles) and energy $e = E(t)/E_{1960}$ (triangles). Value added and capital are adjusted for inflation. Output in the base year, Y_{1960}, is 2263 billion $US\$_{1992}$ in the US, 852.8 billion DM_{1991} in FRG total economy and 453.5 billion DM_{1991} in FRG industry. (The empirical data and their theoretical reproduction with the production function from Eq. (3.42) are taken from [87]. The theory follows in Sect. 3.3)

addition to this almost halved demand for capital goods on the part of investors, there was weakened demand for consumer goods on the part of consumers, because part of their purchasing power had been siphoned off "by the oil sheikhs" through the increase in the price of petrol and heating oil. Following the lower demand, the value

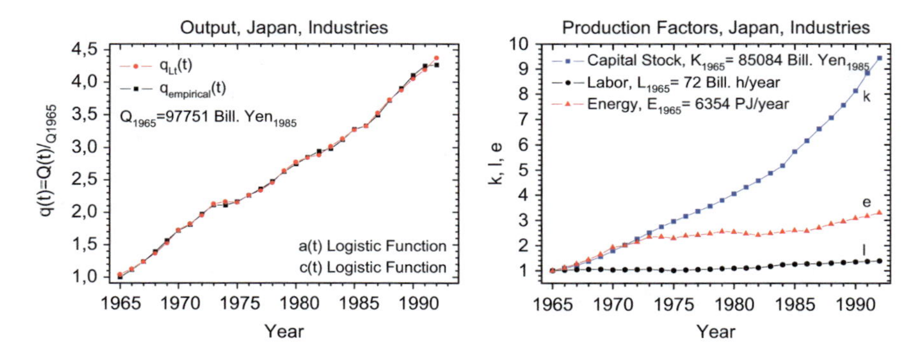

Fig. 3.3 Economic growth in the Japanese "Industries" sector, which produces about 90% of Japan's GDP [2]. Left: Empirical (squares) and theoretical (circles) value added $Q(t)/Q_{1965} \equiv Y(t)/Y_{1965}$, normalized to the base year 1965. Right: Empirical factors of production normalized to the base year capital k, labor l and energy e

added of the overall economy, the GDP, as well as that of the industry was reduced by lower utilization of the production facilities. Correspondingly, less energy was needed and used. Overall, the energy price aggregated from the prices of coal, oil and gas and their cost shares had risen from DM 61.45 per tonne of hard coal units (tce) in 1973 to DM 100.91 per tce in 1975 [88].

Between 1975 and 1979, the price of oil remained stable. OPEC countries did not want to further damage the world economy on which they themselves depended. In addition, nuclear energy began to displace oil from electricity production, and non-OPEC countries such as Britain and Norway developed the oil fields of the North Sea. After the shock wore off, the economy regained momentum even at aggregate energy prices that had risen by about 50%.

Then Saddam Hussein's Iraq invaded revolutionary Iran. This war between two major oil suppliers drastically reduced their oil supply on the world market, and the price of oil shot up, nearly doubling, between 1979 and 1981 to 106 US$$_{2014}$This second oil price shock put the next setback on economic growth in the market economies.

After the end of the Iraq-Iran war, there was an oversupply of oil on the world market. The oil price crashed to 36 US$$_{2014}$ by 1985. For the oil-exporting countries, this was the first *negative* oil price shock. In the oil-importing countries, the economy grew again until, from about the beginning of the twenty-first century, there were the wild price rises and falls shown in Fig. 3.1. The collapse and recovery of economic growth between 2007 and 2010 in Fig. 3.2 are related to the aforementioned mortgage market crisis that began in 2007 because of market poisoning by junk mortgages on U.S. real estate. The crisis erupted openly when the Lehman Brothers investment bank collapsed on 15 September 2008, giving the twenty-first century its first global economic crisis. The bursting of the housing bubble in the US is also associated with the fact that heavily indebted homeowners in suburban

America were no longer able to pay their mortgages because of the sharp rise in oil prices since 2005 and the associated increase in the cost of driving to work [90].

In July 2014, the second negative oil price shock set in with the price of a barrel of crude oil crashing into the US$30–50 range. With peak oil prices prevailing before, the production of oil and gas from non-conventional sources such as oil sands and tar shale had become economical, the US had gone from being an oil importer to being self-sufficient again, and many OPEC countries wanted to reduce their accumulated debt through unlimited oil production and sales. This has been good for the economies of the highly industrialised countries. Problems arose and still arise for oil exporters like Russia, which is painfully reminded of the collapse of the Soviet Union.

The Soviet Union (USSR) had not been a member of OPEC. But as one of the major oil exporters, it benefited enormously from high oil prices on the world market after 1973. While the Western market economies suffered from the first two oil price explosions until 1981, the Warsaw Pact countries enjoyed the low oil prices that Soviet planners granted to both the USSR and its socialist brother states. As a result, East Germany remained virtually unaffected by the two oil price shocks, while West Germany slid into the two recessions shown in Fig. 3.2. At the time, a member of the SED Politburo boasted that now, after all, the superiority of socialism had been proven. In 1979, the leadership of the Soviet Union, chaired by Leonid Brezhnev, General Secretary of the Communist Party of the Soviet Union (CPSU), felt strong enough to occupy Afghanistan and begin an arms race. Naval forces were strengthened to match the U.S. in mastery of the seas, and SS-20 missiles were positioned against Western Europe. But the crash in oil prices after 1981 severely curtailed the flow of petrodollars to the Soviet Union. It was no longer possible to finance the high imports of much-needed consumer goods. Investment of the country's resources had favored heavy industry, while the consumer goods industry, inefficiently organized by the planning bureaucracy, had been neglected and was producing too little. The new General Secretary of the CPSU, Mikhail Gorbachev, recognized the need for change. He introduced the perestroika and glasnost reforms, but it was too late. After the collapse of the Soviet Union in 1991, Russia and the other successor states moved to a market economy—but without the proper legal framework and institutions needed to enforce it. That was one reason for the former superpower's decline in the decade between 1990 and 2000. Another was the low price of oil during the 1990s. When the price of oil picked up towards the end of the twentieth century, rising to 116 $US\$_{2014}$ in 2011, Vladimir Putin came to power in Russia. He became acting president of the Russian Federation on Dec. 31, 1999, won presidential elections in 2000 and 2004, and would have been re-elected in 2008 had not the constitution prohibited a third term. Instead, he was appointed prime minister by his successor, Dmitri Medvedev. After a change in the law that extended the presidential term from 4 to 6 years, Putin won the next presidential election in March 2012. Analysts attribute Putin's popularity in large part to the uplift in the country's standard of living thanks to high Russian revenues from exports of oil and gas during his reign. Since July 2014, however, the price of oil has plummeted even more dramatically than it did after 1981, and this, combined with US and EU sanctions over the Ukraine conflict, has brought Russia economic hardship, but—perhaps precisely because of the Western sanctions—has not (yet) damaged Putin's popular standing. He was re-elected in March 2018.

The previous energy price fluctuations had economic policy reasons. The changes in energy use and economic growth that occurred in their wake are based on bidirectional causality between the two: (1) If, e.g. in the case of an increase in energy prices or a decline in capacity utilisation due to strikes, less energy is demanded and fed into the machinery of the capital stock, fewer goods and services are produced—value added falls. (2) If, e.g. due to the bursting of a speculative bubble with a subsequent stock market crash, the demand for goods and services declines, fewer machines run for their production; only less energy is then required for their operation. If demand recovers so that energy can once again drive machinery, possibly to full capacity, production increases and the economy grows again.

Figures 3.2 and 3.3 show this relationship between energy and economic growth also in the theoretical reproduction of the observed economic development, together with the associated, empirically given factors of production capital K, labour L and energy E. In it, the output Y of goods and services is the annual value added of an economic system, i.e. the gross domestic product (GDP), or a part of it if only one sector of the economy is considered. In the U.S. and Germany, output fluctuates at the rate of energy input, in Japan it kinks like energy input between 1973 and 1975, and overall it grows roughly like capital. Human labor L grows in the U.S., declines in Germany, and remains nearly constant in Japan. The different trends in labor input are almost not noticeable in differences in the growth curves.

3.3 Growth Theory

Production functions like the *LinEx function* from Eq. (3.42) used to calculate theoretical value added in Figs. 3.2 and 3.3, are used to mathematically describe value generation and observed economic growth. They are a standard tool in production and growth theory.

Admittedly, they are viewed sceptically by those economists who either prefer, within the framework of evolutionary economics, to point to the economic significance of thermodynamics by taking a qualitative view of production processes in the economy and "non-anthropogenically manipulated nature", [91] or for the study of economic activities and the associated CO_2 emissions prefer what they call a "*parsimonious*" account and model the production capabilities of individual firms and sectors of the economy by following the Second Law of Thermodynamics in the context of input-output analysis [19]. But for *quantitative* macroeconomic analysis of the driving forces and constraining limits of production and economic growth, production functions are an important tool and are used by many researchers.

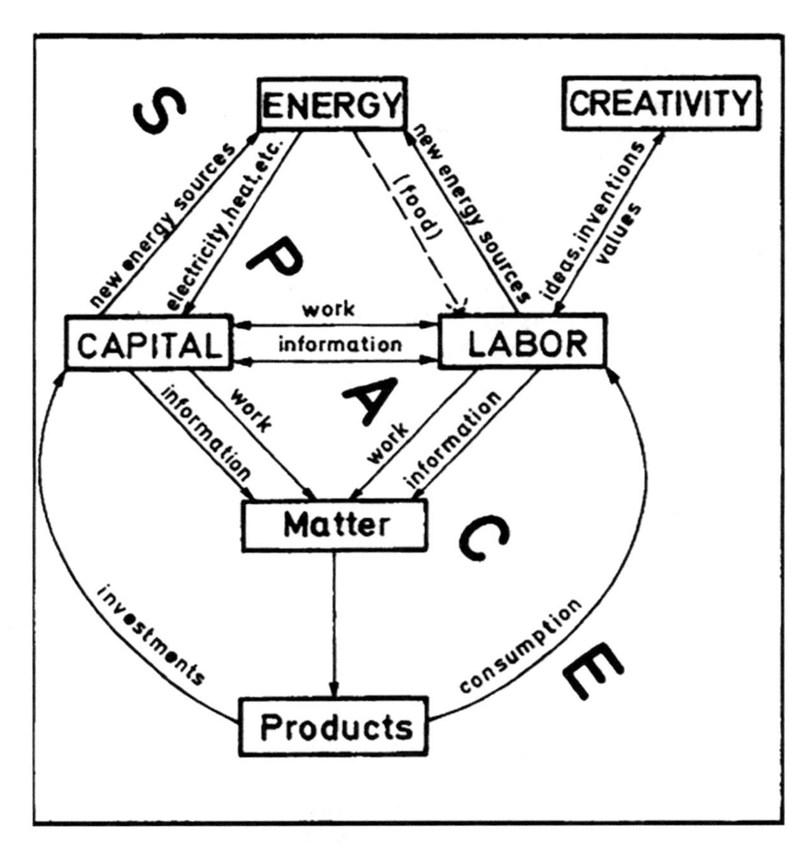

Fig. 3.4 Interaction of the production factors capital, labour and energy with creativity in the space of the economic system [2]

3.3.1 Capital, Labour, Energy and Creativity

As in physics, the mathematical description of a system in economics is based on a system model. Production functions represent macroeconomic models. Figure 3.4 illustrates our model. It sketches the production of an industrial economy by work performance and information processing.

Capital here refers to the capital stock reported by national accounts in inflation-adjusted monetary units. The capital stock consists of all energy conversion devices and information processors and all buildings and installations necessary for their protection and operation.[7] Human labour manipulates the capital stock. It is

[7]The tools and equipment moved by pure muscle power, which are not covered by this concept of capital, make up only a negligible part of the capital of an industrial economy.

measured by national labour market statistics in hours worked per year. Energy activates the capital stock. National energy balances measure it in units of petajoules (PJ); also in use are measures such as tons of coal equivalent (tce) per year, tons of oil equivalent (toe) per year, and other energy units given in Appendix B.1. (The interrupted "food" arrow from "energy" to "labor" indicates that human food is not counted in energy as a factor of production). "Creativity" denotes the influence of human ideas, inventions, and value decisions that shape economic development beyond capital, labor energy.

Price signals from supply and demand link the productive physical substructure of the economy outlined in Fig. 3.4 to the superstructure of the market in which the factors of production and economic goods are traded. This determines the exogenously specified quantities of the factors of production capital, labor and energy used in our model.

Materials, Fig. 3.4 grouped under "matter", are transitory items in the value added balance.[8] They are the passive partners in the production process, in which capital, labour and energy arrange their atoms in such a way that they meet the requirements of the respective product. By means of recycling, using sufficiently large quantities of the three factors of production, the materials can be fed back into the production process again and again at low losses as soon as the economic life of the products formed from them has expired and they have become scrap. As long as there is sufficient energy and the average recycling frequency of the materials is smaller than the inverse average economic lifetime of the products, economic growth need not therefore come up against material-related growth limits. A comprehensive, exergetically based availability analysis of the materials indispensable for modern industrial economies is given in [93].

Information, important in the economy, is linked to and interwoven with the factors of production. Information processing, which in the service sector controls human hands and speech, and which in the industrial sector directs the flow of energy to matter according to the blueprints of designers so that it forms products from raw materials, takes place in the information processors of the human brain and of the capital stock. The simplest information processing consists in opening or closing a switch for energy flows. Information transport is always tied to energy flows, be it electromagnetic waves on which information is imprinted digitally or analogously, be it pulses of electrical currents in conductors or semiconductors that can represent even the smallest unit of information, the bit, or be it the kinetic energy of vehicles that transport newspapers and books. In information storage, printing ink on paper is increasingly being replaced by burning CDs with lasers, electric currents in circuits, magnetization of hard disks, and so on. These stores of information, if

[8]Materials appear as intermediate consumption in the gross output, but not in the gross domestic product, or its parts, which is what we are concerned with here. In the German manufacturing sector, the gross output in 1989 was 2.7 times the contribution of this economic sector to GDP [94].

used in production, are parts of the capital stock.[9] The *creation of* information and knowledge, on the other hand, is a gift of human creativity.

Losses of traditional consumption and investment opportunities amounting to 0.8–6% of German GDP were calculated in various scenarios as a consequence of combating emissions of SO_2, NO_x and CO_2 by means of rational energy use and photovoltaic feed-in tariffs [96]. Nevertheless, for the sake of simplicity, we limit ourselves here to the calculation of GDP, which measures all economic activities valued in monetary terms, i.e. also those that serve to remedy damage caused by economic activities.

Sidenote

Energy theories of value attribute the paramount importance for everything to energy. In labor theories of value, everything is due to human labor. Time scales and system boundaries are decisive. On a time scale since the formation of Earth 4 billion years ago, one may attribute everything to the irradiated solar energy, provided one believes that other impulses "from outside" were not necessary for the unfolding of life and civilization. On a time scale from the beginning of the Neolithic Revolution to the middle of the nineteenth century, and with a system boundary that leaves energy out of the equation because energy had not yet been properly understood and conceptualized physically, one could alternatively see man's labor and technical creativity as the factors that created everything. However, since in highly industrialized countries the produced means of production *capital stock* has grown to the (multiple) size of the gross domestic product and the energy slave figures given in Sect. 2.3.1 far exceed the population figures, capital, labour, energy, and technical creativity can no longer be overlooked as the determining factors of production.

However, according to a utopian vision of digitalisation, energy could perhaps be understood as the only factor of production at some point. That is, if the First Evolutionary Principle of the Factors of Production becomes effective. It states: With growing industrialisation and automation, the production factors capital and labour converge in the production factor energy. This means: In the course of economic development, energy first expands the effectiveness of capital and labour and then increasingly substitutes these two. After the substitution of labor by energy and capital in rationalization measures, the substitution of the factor capital becomes clear by considering the following limiting state. It is the state of the fully automated, computer-controlled factory, regenerating itself in recycling processes of obsolete equipment, into which, in order to maintain uninterrupted production, only energy has to be fed from outside, in addition to the raw materials (from scrapped consumer and capital goods) which can be used again and again. The factor of human labour is completely eliminated, and the factor of capital, which is given with the factory, loses more and more importance in comparison with the factor of energy, if one does not formally attribute the capital goods produced by the factory to the action of

[9] Genetic information stores may also one day become part of the capital stock with further biotechnological progress.

capital (quasi speaking of a miraculous increase of capital), but sees them as having arisen technically-causally through the action of energy [38, 116].

3.3.2 Production Modelling

For further, quantitative analysis of economic growth, we use production functions. Mathematical assumptions and techno-economic ideas lead to orthodox-neoclassical and alternative versions of these functions. The mathematical assumptions are essentially the same. But the path forks when it comes to the techno-economic ideas.

Before the oil price shocks, orthodox economics considered only capital K and labour L as the factors of production to be taken into account as independent variables in the neoclassical production function $A(t)f(K, L)$ when calculating the value added Y_t at time t. A(t) is called "technological progress". It measures technological innovation. And f(K, L) is a twice continously differentiable function of its arguments.

In response to the oil price shocks, economists such as Hudson [97], Jorgenson [98, 100], and Wood [99] introduced energy as a third factor of production in the production function. They were followed by Nordhaus when he studied economic problems of climate change [101]. The production function extended by the factor energy is then $A(t)f_E(K, L, E)$. According to neoclassics, however, in it the weights of the factors of *production*—the *output elasticities of* capital, labour and energy defined in Eq. (3.4)—are predetermined by the factor shares in total factor costs. This *cost-share theorem* is fundamental to neoclassical equilibrium economics and is essentially based on the reasoning that resolves the aforementioned diamond-and-water value paradox (for areas with abundant water). However, its mathematical derivation from maximizing profit (or overall welfare) overlooks *technological constraints* [2, 102, 124]. Section 3.4 addresses this. Since in highly industrialized countries the factor shares of total costs are about 25% for capital, 70% for human labor, and only 5% for energy,[10] neoclassical economists view the contribution of energy to economic growth as marginal at best. This economic disdain for energy is, on the one hand, thermodynamically questionable and, on the other hand, it causes problems for neoclassical growth theory with empiricism.

For example, the econometrician Edward F. Denison argued in a critical discussion of Dale W. Jorgenson's thesis that the decline of GDP in the USA during the years 1973–1975 had something to do with the simultaneous oil price explosion as follows [103]: If one wants to consider energy as a factor of production in its own right and not as a product of capital and land, then one must bear in mind that energy costs in the industrial sector of the USA account for less than 5% of the total factor costs as well as of the value added in this sector. Therefore, the empirically observed decline in energy use of 7.3% in the US industrial sector between 1973 and 1975

[10] OECD values are given in Sect. 6.1.2.

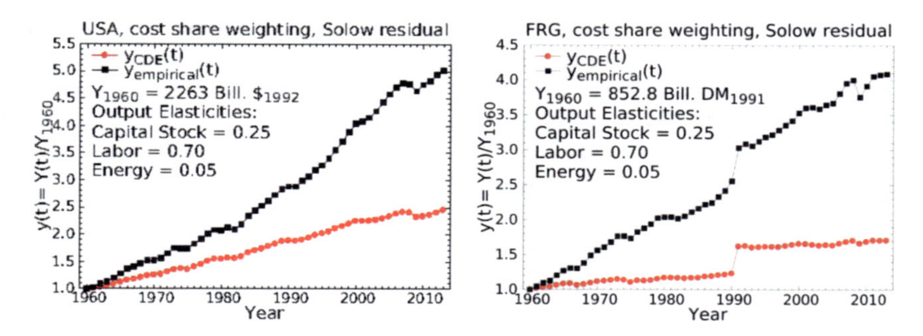

Fig. 3.5 Empirical value added (squares) and theoretically calculated (circles) in the USA and the FRG with the weighting of capital, labour and energy with their cost shares [87]. The theoretical value added is calculated using the energy-dependent Cobb-Douglas function, given by Eq. (3.31)

cannot be related to the observed decline in industrial production of 5.3%[11] [In the original text: "Energy gets about 5 percent of the total input weight in the business sector … the value of primary energy used by nonresidential business can be put at $42 billion in 1975, which was 4.6 percent of a $916 billion nonresidential business national income …. If … the weight of energy is 5 percent, a 1-percent reduction in energy consumption with no change in labor and capital would reduce output by 0.05 percent"].[12]

The inadequate consideration of energy as a factor of production in standard economics had long before been criticised by Tryon and then, at the beginning of the first energy crisis, by Binswanger and Ledergerber. Tryon declared in 1927: "Anything as important in industrial life as power deserves more attention than it has yet received from economists …. A theory of production that will really explain how wealth is produced must analyze the contribution of the element energy" [105]. Binswanger and Ledergerber [106] stated in 1974: "The crucial error of traditional economics (liberal and socialist in character!) is the disregard of energy as a factor of production."

If one weights capital, labour and energy with their above-mentioned cost shares according to neoclassical rules and puts the corresponding output elasticities into the energy-dependent version of the Cobb-Douglas production function (3.31) often used in neoclassics, then one obtains the large deviations of the calculated from the observed economic growth shown in Fig. 3.5. Already the Nobel Prize winner in

[11] The growth of value added in the Industries sector of the US between 1960 and 1978, including the decline in energy use and value added between 1973 and 1975, is theoretically reproduced in [104].

[12] In the industrial sector "goods producing industry" of the old FRG the factor costs in the years 1970 and 1981 amounted to 81 and 156 billion DM for capital, 213 and 258 billion DM for labour, and 11 and 30 billion DM for primary energy (DM figures adjusted for inflation, value 1970). This means: in 1970, when the oil price had its long-term minimum, the share of energy costs in the sum of factor costs in Germany was 3.5%, and in 1981, at the oil price maximum of the twentieth century, industrial energy costs accounted for 7% of total costs [88].

economics Robert M. Solow had found in analyses of the economic growth of the USA during the first half of the twentieth century, when weighting capital and labour with their cost shares, a theoretical economic growth that is significantly below the empirical one [107]. This discrepancy is the notorious *Solow residual.* For its formal elimination, one assumes such a strong growth of the technical progress function $A(t)$ that the discrepancy becomes as small as possible. Thus, the contribution to value added and its growth not explained by the neoclassically weighted factors of production is attributed to the action of *technical progress.* This all-dominant variable, which stands for what is not understood, is sometimes called the "Holy Grail of Economics" or "Manna from Heaven". Solow himself said [108] that the dominant role of technological progress "has led to a criticism of the neoclassical model: it is a theory of growth that leaves the main factor in economic growth unexplained."

The fundamental mathematical properties of production functions assumed by neoclassical economics and also by us are similar to those of physical state functions. They suggest an alternative procedure for determining the economic weights of capital K, labour L and energy E, which replaces the weighting according to factor cost shares. It makes use of the property of state functions to depend only on the current variables of the system and not on the history of the system that led to the current variables.

Production functions of the general form

$$Y = Y(K(t), L(t), E(t); t), \tag{3.1}$$

in which the weighting of the factors is still completely open, model the sum of all entrepreneurial decisions in an economy as the market-influenced decision by a "representative entrepreneur" about the quantities of capital $K(t)$, labour $L(t)$ and energy $E(t)$ to be used at a time t. These decisions are determined by the assessments of the market opportunities of his products.

In the days of the advertising pillar, the entrepreneur relied on intuition and experience regarding customer wishes, and he used posters and advertisements to promote his products. Today, the large Internet corporations determine behavioral patterns and consumer preferences from the personal data with which their customers pay for their services. This information is sold to companies, which use it to determine the production of their goods and services and the corresponding demand for the factors of production. Then they buy targeted advertising on the Internet. The addressees of the advertisements are calculated from the user profiles by the servers' algorithms. Those who do not care about their privacy can see this as a highly efficient global market mechanism.

The ultimate origin and recipient of information are the brain and nervous system. The first push for the mass production of the commodity information came from the printing press. The mass distribution of this commodity continued long after Gutenberg invented printing by horse and rider. Today, as the transistors of computers, smartphones, televisions, and radios convert thoughts into texts and data and send and receive them, often illustrated and multipliable at will, information has become a mass commodity. For private individuals, it is apparently available for free, while entrepreneurs dig deep into their pockets for information to make optimal investment decisions.

The representative entrepreneur invests in the capital stock $K(t)$ taking into account the desired degree of automation, ρ, and in anticipation of a degree of capacity utilization, η. Section 3.4 discusses this in more detail. According to the chosen ρ and η, the entrepreneurial demands for labour $L(t)$ and energy $E(t)$ result. In this sense, capital, labour and energy are the independent variables of the production function of an industrial economy.

The production function (3.1) *implicitly* depends on time via the production factors $K(t)$, $L(t)$ and $E(t)$. The *explicit* time dependence takes into account the effect of "creativity", i.e. the influence of human ideas, inventions and value decisions on economic activity. This general form of production function covers both the neoclassical, orthodox approach extended to include energy, when the factors of production are simply weighted by their cost shares, and alternative production functions. In Sect. 3.3.3, the latter are derived from output elasticities that satisfy a system of differential equations corresponding to Maxwell's relations of thermodynamics and its techno-economic asymptotic boundary conditions. **These boundary conditions are on the signpost that indicates the fork in the methodological road to "alternative output elasticities and production functions". They are called *asymptotic* because they represent limiting cases of economic development.**

Production functions as state functions presuppose causal relationships between the dependent variable, i.e. value added Y, and the independent variables, the factors of production K, L, E. Neoclassical theory takes them as given. We justify it from the aggregation of value added and factors of production into units of work performance and information processing introduced in [104], and presented in more detail in Appendix C. Since work performance and information processing in the energy-activated capital stock are subject to the causal laws of technology, and the same applies to the proper handling of machinery and equipment by humans, value added, the result of the interaction of the three factors, depends causally on them. However, the technical measurement rules for capital and value added given in Appendix C are complex and have not yet been realized. Therefore, for the time being, one must rely on the monetary valuations of capital stock and value added published in the national accounts. These can be assumed to be proportional to the technical measures at some point in time. If the monetary valuations change over time, the corresponding proportionality factors change. This then contributes to the explicit time dependence of the production function (3.1).

Aggregation into units of work performance and information processing also addresses, at least conceptually, the postulation raised in the so-called "Cambridge Controversy" between economists from Cambridge, UK, and Cambridge, Mass. to express value added and factors of production in physical units in order for macroeconomic production functions to make sense [109–111].

Sidenote

For the mathematical equal treatment of capital, labour and energy in the production function, it is crucial that these three factors are the variables freely chosen by the entrepreneur: The representative entrepreneur can increase the capital stock by investing or decrease it by not investing in replacements and/or by shifting production abroad; within the technological constraints discussed in Sect. 3.4, he chooses the hours of labour and amounts of energy to be used per time interval according to targeted levels of utilisation and automation of the capital stock. Differences between *stocks and flows*, which *are* sometimes emphasized by heterodox economists, do not play a role.

Moreover, the difference between stock and flow quantities blurs with the extension of time scales: Within a year, the capital stock and the available labour may appear as stock quantities, while the primary energy used is converted into waste heat, which ends up in useless ambient heat—unless it is further used in rational energy use through heat recovery, as calculated in Sect. 2.3.3. However, on time scales ranging from the average lifetime of capital goods to the duration of human working capacity, capital and labour also wear out and become flow variables.

3.3.3 Productive Powers

The growth rates of the gross domestic product or the value added of individual economic sectors are regarded as indicators of the dynamics and strength of an economy or its sectors. They sometimes determine the weal and woe of governments. How do capital, labour, energy and innovations contribute? Also against the background of the coupling of energy conversion to emissions described in Chap. 1 on the one hand and the manifold services of energy slaves listed in Chap. 2 on the other hand, a quantitative determination of the factor contributions to economic growth is necessary for the assessment of the possibilities of sustainable economic development. In other words, we are looking for the weights with which the growth rates of the capital stock, of the hours worked per year and of the quantities of primary energy used per year, in combination with innovations, contribute to the growth of value added. We refer to these weights as α for capital, β for labor, and γ for energy. Economics calls them *output elasticities*. We also speak of *productive powers*, because they measure the productive powers of the factors K, L and E. We denote the productive power of innovations brought about by human creativity, measured by changes in efficiency parameters over time, by δ.

In the following, the mathematical theory is outlined which allows first to calculate output elasticities and then production functions depending on capital, labour and energy and thus to determine the corresponding numerical values of α, β, γ and δ from observed economic growth, i.e. the empirically given time series of the production factors and the value added of an economic system. Readers to whom these numerical values are more important for the time being than the mathematical apparatus behind them can jump from here directly to the paragraph following Eq. (3.45).

The calculation of output elasticities and production functions begins by considering an infinitesimally small change in value added, dY. Such a change is equal to the total differential

$$\mathrm{d}Y(K, L, E; t) = \frac{\partial Y}{\partial K}\mathrm{d}K + \frac{\partial Y}{\partial L}\mathrm{d}L + \frac{\partial Y}{\partial E}\mathrm{d}E + \frac{\partial Y}{\partial t}\mathrm{d}t \tag{3.2}$$

of the production function $Y(K, L, E; t)$ describing the output (value added) of the economic system (For simplicity, the time dependence of K, L, E is no longer indicated in the notation). Dividing $\mathrm{d}Y$ by the production function $Y(K, L, E; t)$, we obtain the growth rate of value added. Its dependence on the growth rates of the factors of production and the innovations over time is described by the growth equation

$$\frac{\mathrm{d}Y}{Y} = \alpha\frac{\mathrm{d}K}{K} + \beta\frac{\mathrm{d}L}{L} + \gamma\frac{\mathrm{d}E}{E} + \delta\frac{\mathrm{d}t}{\Delta t}. \tag{3.3}$$

The weights of the growth rates of capital, α, labour, β, and energy, γ, i.e. the output elasticities of the factors, indicate, roughly speaking, the percentage change in value added for a 1% change in the factors. In this sense, they measure the productive powers of the factors of production. From the comparison of (3.3) with (3.2) follows their definition:

$$\alpha \equiv \frac{K}{Y}\frac{\partial Y}{\partial K}, \quad \beta \equiv \frac{L}{Y}\frac{\partial Y}{\partial L}, \quad \gamma \equiv \frac{E}{Y}\frac{\partial Y}{\partial E}; \tag{3.4}$$

furthermore, the comparison shows that

$$\delta \equiv \frac{\Delta t}{Y}\frac{\partial Y}{\partial t}. \tag{3.5}$$

We denote δ, analogous to α, β, γ, as the output elasticity of creativity. We choose $\Delta t = |t - t_0|$, where t_0 is any base year with factor inputs K_0, L_0, E_0 (This choice is likely to emphasize long-run effects of creativity more than another choice, e.g. $\Delta t = 1$ year).

As a state function, $Y(K, L, E; t)$ must be twice continuously differentiable, i.e. its mixed second derivatives with respect to K, L, E must be equal. From this follow the (Cauchy) integrability conditions of the growth Eq. (3.3):

$$L\frac{\partial \alpha}{\partial L} = K\frac{\partial \beta}{\partial K}, \quad E\frac{\partial \beta}{\partial E} = L\frac{\partial \gamma}{\partial L}, \quad K\frac{\partial \gamma}{\partial K} = E\frac{\partial \alpha}{\partial E}. \tag{3.6}$$

They correspond to the Maxwell relations of thermodynamics, and like these they are obtained from the total differential of the state function, in this case from Eq. (3.2).

We integrate the growth Eq. (3.3) at a fixed time t at which the factors of production $K = K(t)$, $L = L(t)$, $E = E(t)$ act. The integral of the left-hand side from $Y_0(t)$ to $Y(K, L, E; t)$ gives $\ln \frac{Y(K,L,E;t)}{Y_0(t)}$. It is equal to the path integral of the right-hand side,

$$F(K,L,E)_t \equiv \int_{P_0}^{P} \left[\alpha \frac{\mathrm{d}K}{K} + \beta \frac{\mathrm{d}L}{L} + \gamma \frac{\mathrm{d}E}{E} \right] \mathrm{d}s. \tag{3.7}$$

If the output elasticities satisfy the integrability conditions (3.6), this integral can be evaluated along any path s in factor space from a starting point P_0 at (K_0, L_0, E_0) to the end point P at $(K(t), L(t), E(t))$.

A convenient path consists of three straight lines parallel to the Cartesian axes of K, L, E *space*: $P_0 = (K_0, L_0, E_0) \rightarrow P_1 = (K, L_0, E_0) \rightarrow P_2 = (K, L, E_0) \rightarrow P = (K, L, E)$. This gives

$$\begin{aligned} F(K,L,E)_t &= \int_{K_0,L_0,E_0}^{K,L_0,E_0} \alpha(K,L_0,E_0) \frac{\mathrm{d}K}{K} + \int_{K,L_0,E_0}^{K,L,E_0} \beta(K,L,E_0) \frac{\mathrm{d}L}{L} \\ &+ \int_{K,L,E_0}^{K,L,E} \gamma(K,L,E) \frac{\mathrm{d}E}{E}. \end{aligned} \tag{3.8}$$

Since $\ln \frac{Y(K,L,E;t)}{Y_0(t)} = F(K,L,E)_t$, the general production function (3.1) has the form

$$Y(K,L,E;t) = Y_0(t) \exp\{F(K,L,E)_t\}. \tag{3.9}$$

The integration constant $Y_0(t)$ is the monetary value of a basket of goods and services that would be produced at time t using the factors K_0, L_0, and E_0. If creativity is dormant during the time interval $|t - t_0|$, $Y_0(t)$ is also equal to the production function at time t_0. Active creativity, on the other hand, can change Y_0. The same is true for the two technology parameters (a and c) in the LinEx functions calculated in Sect. 3.3.4 and Appendix A.4.

At any given time t, the contributions of the growth rates of K, L, E to the growth rate of value added must add up to 100%, i.e. the output elasticities must satisfy the so-called "relation of constant returns to scale":

$$\alpha + \beta + \gamma = 1. \tag{3.10}$$

This is because if you add an identical production system with the same factor inputs to an existing one, the value added must double. The production function is therefore *linearly homogeneous* (More precisely, it must be true for the production function that $Y(\lambda K, \lambda L, \lambda E; t) = \lambda Y(K, L, E; t)$, for each $\lambda > 0$ and all possible factor combinations. Differentiating this equation by λ according to the chain rule and then setting $\lambda = 1$, we obtain the Euler relation $K\partial Y/\partial K + L\partial Y/\partial L + E\partial Y/\partial E = Y$. Dividing this equation by Y and observing (3.4) yields Eq. (3.10)).

Moreover, the output elasticities must be non-negative. Otherwise, an increase in a production factor would reduce value added—a situation that economic agents avoid. Therefore, the restrictions apply:

$$\alpha \geq 0, \quad \beta \geq 0, \quad \gamma \geq 0. \tag{3.11}$$

Rewriting Eq. (3.10) to $\gamma = 1 - \alpha - \beta$ and substituting this γ into the integrability conditions (3.6), these become

$$K \frac{\partial \alpha}{\partial K} + L \frac{\partial \alpha}{\partial L} + E \frac{\partial \alpha}{\partial E} = 0, \tag{3.12}$$

$$K \frac{\partial \beta}{\partial K} + L \frac{\partial \beta}{\partial L} + E \frac{\partial \beta}{\partial E} = 0, \tag{3.13}$$

$$L \frac{\partial \alpha}{\partial L} = K \frac{\partial \beta}{\partial K}. \tag{3.14}$$

The general solutions of Eqs. (3.12), (3.13), and (3.14) are the output elasticities

$$\alpha = \alpha\left(\frac{L}{K}, \frac{E}{K}\right), \quad \beta = \beta\left(\frac{L}{K}, \frac{E}{K}\right). \tag{3.15}$$

Furthermore, because of Eq. (3.14)

$$\beta = \int^{K} \frac{L}{K'} \frac{\partial \alpha}{\partial L} \mathrm{d}K' + J(L/E). \tag{3.16}$$

$\alpha(L/K, E/K)$, $\beta(L/K, E/K)$ and $J(L/E)$ are continuous differentiable functions of their arguments.

The two equations in (3.15) follow from the theory of partial differential equations. They and Eq. (3.16) can be verified by substituting $\alpha(L/K, E/K)$ and $\beta(L/K, E/K)$ into Eqs. (3.12) and (3.13), as well as $\alpha(L/K, E/K)$ and the β of Eq. (3.16) into Eq. (3.14).

The general form of the twice continuous-differentiable, linear-homogeneous production function corresponding to the output elasticities (3.15) and $\gamma = 1 - \alpha - \beta$ is then

$$Y = E\mathcal{F}\left(\frac{L}{K}, \frac{E}{K}\right). \tag{3.17}$$

If, on the other hand, we write the constant-returns-to-scale relation (3.10) as $\beta = 1 - \alpha - \gamma$ and insert this β into the integrability conditions (3.6), instead of the system of Eqs. (3.12), (3.13), and (3.14) we obtain the equations

$$K \frac{\partial \alpha}{\partial K} + L \frac{\partial \alpha}{\partial L} + E \frac{\partial \alpha}{\partial E} = 0, \tag{3.18}$$

$$K \frac{\partial \gamma}{\partial K} + L \frac{\partial \gamma}{\partial L} + E \frac{\partial \gamma}{\partial E} = 0, \tag{3.19}$$

$$E\frac{\partial\alpha}{\partial E}=K\frac{\partial\gamma}{\partial K}. \tag{3.20}$$

Following the above reasoning again, one finds as general solutions of Eqs. (3.18), (3.19), and (3.20) the output elasticities

$$\alpha=\alpha\left(\frac{L}{K},\frac{E}{K}\right),\quad \gamma=\gamma\left(\frac{L}{K},\frac{E}{K}\right), \tag{3.21}$$

where

$$\gamma=\int^{K}\frac{E}{K'}\frac{\partial\alpha}{\partial E}\mathrm{d}K'+P(E/L) \tag{3.22}$$

and as a form of the general production function

$$Y=L\mathcal{G}\left(\frac{L}{K},\frac{E}{K}\right). \tag{3.23}$$

Correspondingly, combining $\alpha = 1 - \beta - \gamma$ with (3.6) yields a third system of equations:

$$K\frac{\partial\gamma}{\partial K}+L\frac{\partial\gamma}{\partial L}+E\frac{\partial\gamma}{\partial E}=0 \tag{3.24}$$

$$K\frac{\partial\beta}{\partial K}+L\frac{\partial\beta}{\partial L}+E\frac{\partial\beta}{\partial E}=0 \tag{3.25}$$

$$L\frac{\partial\gamma}{\partial L}=E\frac{\partial\beta}{\partial E}. \tag{3.26}$$

Its solutions are

$$\beta=\beta\left(\frac{L}{K},\frac{E}{K}\right),\quad \gamma=\gamma\left(\frac{L}{K},\frac{E}{K}\right), \tag{3.27}$$

with

$$\gamma=\int^{L}\frac{E}{L'}\frac{\partial\beta}{\partial E}\mathrm{d}L'+R(E/K). \tag{3.28}$$

The general production function then has the form

$$Y=K\mathcal{H}\left(\frac{L}{K},\frac{E}{K}\right). \tag{3.29}$$

In Sect. 3.3.5 and Appendix A.4, explicit forms of the output elasticities and the associated production functions (3.17), (3.23), (3.29) are calculated for production systems of different degrees of industrialization and automation from different asymptotic boundary conditions of the partial differential equation systems (3.12), (3.13), (3.14), (3.18), (3.19), (3.20), (3.24), (3.25), and (3.26).

Before that, let us take a look at neoclassical theory.

3.3.4 Neoclassicism

Neoclassical production functions were constructed by their "inventors" according to desired functional properties and were not obtained by integrating the growth Eq. (3.3) with corresponding output elasticities. Textbooks such as [112] refer to the factor-independent constant that takes the place of the integration constant $Y_0(t)$ in the neoclassical production function as the efficiency parameter or level parameter. Among the properties that neoclassical production functions are said to have is that the isoquant connecting the points of equal output Y when two factors of production, e.g. K and L, vary is always below the chord connecting two of their points; it is also said to be "convex to the origin" of the $K - L$-axis axes of coordinates in order to describe an efficient production system [112]. It is also required that each of the second derivatives of the production function with respect to the individual factors of production is negative, so that the marginal productivity $\partial Y/\partial X$ of factor X decreases as X increases, and the law of diminishing returns formulated in Sect. 3.3.5 is already modelled by the second derivatives (curvatures) of the production function.

Based on the growth Eq. (3.3) and the definition (3.4) of output elasticities, it follows after a short calculation that the conditions $\partial^2 Y/\partial K^2 < 0$, $\partial^2 Y/\partial L^2 < 0$, $\partial^2 Y/\partial E^2 < 0$ are satisfied if

$$\alpha - \alpha^2 > K\frac{\partial \alpha}{\partial K}, \quad \beta - \beta^2 > L\frac{\partial \beta}{\partial L}, \quad \gamma - \gamma^2 > E\frac{\partial \gamma}{\partial E}. \tag{3.30}$$

The simplest production function with the required mathematical properties that is frequently used in textbook economics is the Cobb-Douglas function of the factors capital and labour proposed by C.W. Cobb and P.H. Douglas in the 1920s. If one extends it by the factor energy E, it has with the constant output elasticities α_0, β_0 and $\gamma_0 = 1 - \alpha_0 - \beta_0$ the form

$$Y_{CDE}(K, L, E; t) = Y_0(t)\left(\frac{K}{K_0}\right)^{\alpha_0}\left(\frac{L}{L_0}\right)^{\beta_0}\left(\frac{E}{E_0}\right)^{1-\alpha_0-\beta_0}. \tag{3.31}$$

It also results from the trivial, i.e. constant, solutions of the differential Eqs. (3.12), (3.13), and (3.14), if one inserts these solutions into Eqs. (3.8) and (3.9). With the assumption, characteristic of neoclassics, that output elasticities are equal to factor costs, shares i.e. that approximately $\alpha_0 = 0,25$, $\beta_0 = 0,7$ $\gamma_0 = 0,05$, one obtains (with fixed $Y_0(t)$) the theoretical economic growth of Fig. 3.5, which differs from empirical growth by the Solow residuals.

In order to eliminate the Solow residuals, textbook economics replaces $Y_0(t)$ with the previously mentioned function $A(t)$, which is allowed to grow with time *t to* such an extent, e.g. exponentially, that the theoretical growth curve deviates as little as possible from empirical growth. The high contribution of this function $A(t)$ to economic growth is interpreted as the contribution of scientific and technical progress.

Often, standard economics also considers production functions with *constant elasticity of substitution,* the so-called CES functions. They were introduced by Arrow et al. [113] and extended to more than two factors of production by Uzawa [114]. The linear homogeneous CES function in *K*, *L*, *E* has the form [115].

$$Y_{CES}(K, L, E; t) = A(t)\left[aK^{-\rho} + bL^{-\rho} + (1 - a - b)E^{-\rho}\right]^{-1/\rho}. \tag{3.32}$$

The parameters a and b are non-negative, and the constant $\rho \equiv 1/\sigma - 1$, which is determined by the constant elasticity of substitution σ, must be greater than -1. The CES function can easily be put into the form of Eq. (3.17). It satisfies the growth Eq. (3.3) as well as Eq. (3.12), (3.13), and (3.14). Its output elasticities are obtained according to Eq. (3.4) as follows

$$\alpha_{CES} = a\left(\frac{Y_{CES}}{Y_0(t)K}\right)^{\rho}, \quad \beta_{CES} = b\left(\frac{Y_{CES}}{Y_0(t)L}\right)^{\rho}; \tag{3.33}$$

γ_{CES} follows from (3.10). In the limiting case $\sigma \to 1$, when $\rho \to 0$, the CES function (3.32) becomes the Cobb-Douglas function (3.31). This is most easily seen in the output elasticities (3.33), which become constants in this limiting case.

Translog functions approximate the general production function by Taylor evolutions up to second order in $\ln \frac{L/L_0}{K/K_0}$ and $\ln \frac{E/E_0}{K/K_0}$ [89].

Supplement: Variable Elasticities of Substitution

The (Hicks, or direct) elasticity of substitution σ_{ij} of the factor of production x_i by the factor x_j is defined as

$$\sigma_{ij} \equiv -\frac{\mathrm{d}(x_i/x_j)}{(x_i/x_j)} \cdot \left[\frac{\mathrm{d}\left(\frac{\partial y/\partial x_i}{\partial y/\partial x_j}\right)}{\left(\frac{\partial y/\partial x_i}{\partial y/\partial x_j}\right)}\right]^{-1}. \tag{3.34}$$

It is the ratio of the relative change in factor quotients to the relative change in marginal productivity quotients when only the factors x_i and x_j change and all other factors remain constant.

In the three-factor model with $(x_1, x_2, x_3) = (K, L, E)$, one can express the elasticities of substitution by the elasticities of production α, β and γ. After some algebraic manipulations, one obtains [115]

$$\sigma_{KL} = \frac{-(\alpha+\beta)\alpha\beta}{\beta^2(K\partial\alpha/\partial K-\alpha)+\alpha^2(L\partial\beta/\partial L-\beta)-2\alpha\beta L\partial\alpha/\partial L}, \tag{3.35}$$

$$\sigma_{KE} = \frac{-(\alpha+\gamma)\alpha\gamma}{\gamma^2(K\partial\alpha/\partial K-\alpha)+\alpha^2(E\partial\gamma/\partial E-\gamma)-2\alpha\gamma E\partial\alpha/\partial E}, \tag{3.36}$$

$$\sigma_{LE} = \frac{-(\beta+\gamma)\beta\gamma}{\gamma^2(L\partial\beta/\partial L-\beta)+\beta^2(E\partial\gamma/\partial E-\gamma)-2\beta\gamma E\partial\beta/\partial E}. \tag{3.37}$$

3.3.5 From the Law of Diminishing Returns to LinEx Functions

The price and quantity of a factor of production determine its share in total factor costs. We have seen that with the previous prices for energy and labour the neoclassical assumption "output elasticity = factor cost share" leads to production functions with large Solow residuals and low explanatory power. The function $A(t)$ introduced for the stopgap "technical progress" has only the meaning of a general level parameter without further technical-economic specification. Thus, the economic fluctuations in reaction to oil price shocks remain economically incomprehensible, and thermodynamically incomprehensible is the low significance thus assigned to energy in highly industrialized economies such as the USA, Japan and Germany.[13]

Alternatively, techno-economic relationships point the way to computing output elasticities from the differential equation systems (3.12), (3.13), and (3.14) or (3.18), (3.19), and (3.20) or (3.24), (3.25), and (3.26).

The guide to the boundary conditions needed to solve these differential equations is the law of diminishing returns. This "famous techno-economic relation" [33], is formulated by Samuelson [33, Vol. I, p. 44 ff] as follows:

> For a given state of technology, the additional input of a factor, holding the other factor inputs constant, causes an increase in output; however, from a certain point on, the additional output of an additional unit of the variable factor will decrease. This decrease is due to the fact that one unit of increasing factor is combined with smaller and smaller quantities of the fixed factors.

Samuelson emphasizes the fundamental importance of the law of diminishing returns because of empirical facts that seem to contradict this law when only capital and labor are considered as factors of production: In Fig. 37.3 of his textbook [33],

[13]This is also expressed in the language used in the stock market news. There, people talk about energy as the "lubricant"—not fuel—of the economy. Apparently, no one thinks about the fact that drivers cannot get far without regularly refueling with gasoline or diesel, but only occasionally, and increasingly rarely, need to top off transmission fluid. With electric cars, the differences between "refueling" and "lubricating" become even more impressive.

between 1900 and 1970, the ratio of capital to labor, K/L, also called the *depth of production*, increases steadily, while the value added Y grows with the capital stock K *in* such a way that the ratio K/Y, also called the *capital coefficient*, remains about the same. Our Fig. 3.2 continues this trend for the U.S. through 2013 (In Germany's economy as a whole, K/Y grows by a factor of 1.5 between 1960 and 2013, but this growth lags far behind that of the depth of production K/L, which increases sevenfold). So, with the observed trends in the depth of production and the coefficient of capital, is there a violation of the law of diminishing returns? Not at all, says Samuelson, pointing out that "contemporary economic theorists believe that scientific and technical progress in industrialized nations has been and still is the quantitatively most important cause of growth" and that this progress only conceals the law [33, footnote 15, p. 483, Vol. II].

Based on this hint, it was suggested to introduce a production **depth** $K/(E + L)$ **extended** by the production factor energy E, to assume the output elasticity of capital α as proportional to $(E + L)/K$ and to use such a α in the equation for the growth rate dY/Y [116]. Thus began the search for production functions whose output elasticities are the solutions of the differential equation system (3.12), (3.13), and (3.14) together with appropriate techno-economic boundary conditions, thus representing scientific and technological progress as the action of energy and creativity.

To calculate *exact* output elasticities and production functions, one would need to know the exact boundary conditions. According to the theory of partial differential equations, these would be, for the system of Eqs. (3.12), (3.13), and (3.14) at any time t, all values of β on a boundary surface in K, L, E *space* and those of α on a boundary curve (One would need to know the equivalent for the other two systems of differential equations). This information is not available to anyone. In this sense, all boundary conditions are approximations and so are the associated output elasticities and production functions. One cannot know the exact production function of an economic system.

Sidenote

Tintner et al. in one of the first econometric studies that did *not* equate the output elasticities in the energy-dependent Cobb-Douglas function to the factor cost shares obtained an energy output elasticity of more than 30% for the Austrian economy between 1955 and 1972 [117]. Furthermore, fitting the Cobb-Douglas function Y_{CDE} of Eq. (3.31) by minimizing the sum of squared errors (3.44) to the empirical growth in the economic sector "FRG industry" shown in Fig. 3.2 yields the output elasticities shown in the left part of Fig. 3.6, which exceed 50% for energy and fall below 10% for labor. The theoretical reproduction of empirical value added in the right part of the figure is satisfactory between 1960 and 1989, but it becomes deficient after reunification in 1990 (The fits of Y_{CDE} to the overall U.S. and FRG economies between 1960 and 2013 are woefully inadequate. They provide almost smooth theoretical curves that essentially follow the growth of the capital stock and suppress the fluctuations).

Over shorter periods of time, and for an economic sector as strongly dependent on energy as "FRG industry", Y_{CDE} may be sufficient for the reproduction of past

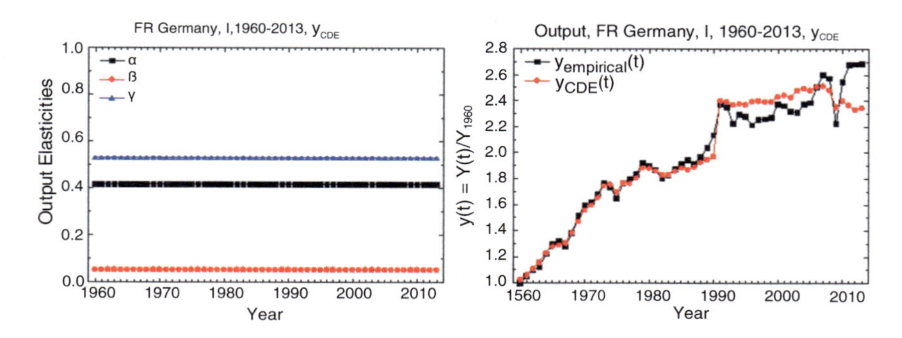

Fig. 3.6 FRG industry 1960–2013. Output elasticities (left) and growth of value added (right) according to the energy-dependent Cobb-Douglas production function of Eq. (3.31) [119]

economic growth *without* equating output elasticities and factor cost shares. But for the estimation of future economic developments the Cobb-Douglas function is not suitable, namely because of the mathematically admissible but thermodynamically impossible (asymptotically) complete mutual substitutability of the factors K, L, E. For the past this substitutability does not play an essential role, because the economic actors know, and have always known from experience, that energy cannot be completely replaced by capital, or vice versa.

Each of the three formally different versions (3.17), (3.23), (3.29) of the general production function belongs to a different set of asymptotic boundary conditions. If one knows such conditions for α and β, one obtains solutions of the system of Eqs. (3.12), (3.13), and (3.14) and special production functions of the type (3.17). From boundary conditions for α and γ and solutions of (3.18), (3.19), and (3.20), one obtains functions of type (3.23). Boundary conditions for β and γ give solutions of (3.24), (3.25), and (3.26) and production functions of type (3.29).

We are looking for asymptotic boundary conditions for three development stages of an economy—highly industrialized, early industrialized and totally digitalized—and we proceed according to the method of "Ockham's Razor" (also: KISS = Keep It Simple, Stupid). In other words: We are looking for the simplest boundary conditions that represent the law of diminishing returns in the three economic development stages. To this end, we construct the simplest output elasticities, each of which satisfies one of the three differential equation systems. The production functions thus calculated all depend linearly on one of the three factors and exponentially on quotients of all three factors. In this sense, they are all members of the family of LinEx functions.

Highly Industrialized Countries of the Present

Today's highly industrialised countries were already encompassed by the Industrial Revolution in the eighteenth and nineteenth centuries. They include the G7 countries of Germany, France, Italy, Japan, Canada, the USA and the United Kingdom of Great Britain and Northern Ireland. In 2006, they accounted for more than 64% of global value added. Since the early twentieth century, their economic development has been shaped by the dynamics of energy use and automation. This suggests the following asymptotic boundary conditions.

1. The contribution α of the growth of the capital stock to the growth of value added should become vanishingly small, according to the law of diminishing returns, when the ratios of labor to capital and energy to capital become vanishingly small.
2. When the capital stock approaches the state of maximum automation characterized by the point $(K_m(t), L_m(t), E_m(t))$ in the factor space, an additional unit of labor L no longer contributes to the growth of value added, so β must disappear. Here $K_m(t)$ is the maximum automated capital stock that, in combination with $L_m(t) \ll L(t)$ and $E_m(t) \equiv c(t)E_0K_m(t)/K_0$, could produce an output Y_t given at time t if no constraints on the volume, mass, and energy requirements of the information processors existed at time t in the finite space of the production system. The technology parameter $c(t)$ measures the energy demand of the fully utilized capital stock present at time t, i.e. $1/c(t)$ is the energy efficiency parameter of the production system at time t.

All changes are considered relative to the base year t_0. Thus, the two asymptotic boundary conditions are as follows

$$\alpha \to 0, \quad \text{If} \quad \frac{L/L_0}{K/K_0} \to 0, \quad \frac{E/E_0}{K/K_0} \to 0, \tag{3.38}$$

$$\beta \to 0, \quad \text{If} \quad K \to K_m(t), \quad E \to E_m \equiv cE_0 \frac{K_m(t)}{K_0}. \tag{3.39}$$

The simplest factor-dependent output elasticities that satisfy these boundary conditions and the differential Eqs. (3.12), (3.13), and (3.14) when we choose $J(L/E) = ac\frac{L/L_0}{E/E_0}$, are, on the one hand

$$\begin{aligned} \alpha &= a\frac{L/L_0 + E/E_0}{K/K_0}, \quad \beta = a\left(c\frac{L/L_0}{E/E_0} - \frac{L/L_0}{K/K_0}\right), \\ \gamma &= 1 - a\frac{E/E_0}{K/K_0} - ac\frac{L/L_0}{E/E_0}, \end{aligned} \tag{3.40}$$

and on the other hand

$$\begin{aligned} \alpha &= a\frac{L/L_0}{K/K_0} + \frac{1}{c}\frac{E/E_0}{K/K_0}, \quad \beta = a\left(c\frac{L/L_0}{E/E_0} - \frac{L/L_0}{K/K_0}\right), \\ \gamma &= 1 - \frac{1}{c}\frac{E/E_0}{K/K_0} - ac\frac{L/L_0}{E/E_0}. \end{aligned} \tag{3.41}$$

The technology parameter a, which complements the energy demand parameter c, measures capital efficiency. It has somewhat different meanings in (3.40) and (3.41): In (3.40) it is the weight with which the ratio of labor *plus* energy to capital contributes to the output elasticity of the capital stock, while in (3.41) parameter a concerns only the ratio of labor to capital.

With these output elasticities, Eqs. (3.8) and (3.9) provide the LinEx production functions

$$Y_{L1}(K, L, E; t) = Y_0(t) \frac{E}{E_0} \exp\left[a\left(2 - \frac{L/L_0 + E/E_0}{K/K_0}\right) + ac\left(\frac{L/L_0}{E/E_0} - 1\right)\right] \tag{3.42}$$

and

$$Y_{L11}(K, L, E; t) = Y_0(t) \frac{E}{E_0} \exp\left[a\left(1 - \frac{L/L_0}{K/K_0}\right) + \frac{1}{c}\left(1 - \frac{E/E_0}{K/K_0}\right) + ac\left(\frac{L/L_0}{E/E_0} - 1\right)\right]. \tag{3.43}$$

Only with these two LinEx functions econometric analyses have been performed so far. For this purpose, time dependencies of the technology parameters were modeled [2, 87, 118] and the $a(t)$ and $c(t)$ were determined by minimizing the sum of squared errors SSE:

$$\text{Minimize} \quad \text{SSE} = \sum_i \left[Y_{\text{empirical}}(t_i) - Y(t_i)\right]^2 \tag{3.44}$$

subject to the constraints

$$\alpha \geq 0, \quad \beta \geq 0, \quad \gamma = 1 - \alpha - \beta \geq 0. \tag{3.45}$$

$Y_{\text{empirical}}(t_i)$ is the empirical value added in the year t_i, and $Y(t_i)$ is the abbreviation for the production function with the production factors $K(t_i)$, $L(t_i)$, $E(t_i)$. Sums are taken over all years t_i from the first to the last year of observation.

The latest numerical calculations for Germany and the USA were carried out by Tobias Winkler [119]. The most important details and results were published in [87]. Figure 3.2 shows the growth of value added in the two countries between 1960 and 2013 calculated with the simplest LinEx function $Y_{L1}(K, L, E; t)$. The growth calculated with $Y_{L11}(K, L, E; t)$ is almost identical. Table 3.1 gives the time averages of the output elasticities of capital, labour, energy and creativity. These differ only slightly for the two LinEx functions.

For the second LinEx function $Y_{L11}(K, L, E; t)$ a higher numerical effort has to be made. Mathematically, it has the advantage of being invariant to transformations of the base year t_0 [87], while for $Y_{L1}(K, L, E; t)$ a noticeable influence of the base year on the results can so far (only) be denied after carrying out the respective numerical calculations.

According to Table 3.1, the averaged output elasticities of labour, $\overline{\beta}$, are smaller than the cost share of labour of around 70% by a factor of about 3 in the overall economies of Germany and the USA and by a factor of 9 in the German industrial

Table 3.1 Temporal mean values of the output elasticities of capital, $\overline{\alpha}$, labour, $\overline{\beta}$, energy, $\overline{\gamma}$, creativity, $\overline{\delta}$, obtained with the LinEx production functions Y_{L1} and Y_{L11}, for the production systems FR Germany total economy (FRG TE), FR Germany industry (FRG I) and USA total economy (USA TE)

	FRG TE	FRG I	USA TE
Y_{L1}			
$\overline{\alpha}$	0.367 ± 0.006	0.284 ± 0.008	0.518 ± 0.023
$\overline{\beta}$	0.188 ± 0.004	0.076 ± 0.008	0.188 ± 0.041
$\overline{\gamma}$	0.445 ± 0.007	0.640 ± 0.011	0.294 ± 0.047
$\overline{\delta}$	0.217 ± 0.006	0.132 ± 0.007	0.200 ± 0.023
R^2	0.999	0.989	0.998
d_W	1.650	1.747	0.715
Y_{L11}			
$\overline{\alpha}$	0.399 ± 0.008	0.221 ± 0.020	0.533 ± 0.016
$\overline{\beta}$	0.236 ± 0.012	0.015 ± 0.009	0.242 ± 0.035
$\overline{\gamma}$	0.365 ± 0.014	0.765 ± 0.022	0.226 ± 0.039
$\overline{\delta}$	0.236 ± 0.032	0.27 ± 0.16	0.168 ± 0.019
R^2	0.999	0.988	0.999
d_W	1.508	1.581	0.762

The observation period covers the years 1960 to 2013. $\overline{R}^2$ and d_W are the corrected multiple coefficient of determination and the Durbin-Watson autocorrelation coefficient

sector ("goods-producing industry"). The averaged output elasticities of energy, $\overline{\gamma}$, on the other hand, exceed the cost share of energy of about 5% by a factor of five to ten. The dominant role of capital in the US is noteworthy, where the mean output elasticity of capital, α, is more than twice the cost share of this factor. The mean value of the contribution of creativity to growth calculated from the time dependence of the technology parameters a and c, $\overline{\delta}$, is significantly smaller than that of energy.

Overall, energy and creativity appear to be the larger and smaller contributors to that part of economic growth that standard economics attributes to scientific and technological progress.

The time dependencies of the technology parameters and the output elasticities are presented in [87].

Sidenote

The neoclassical curvature conditions (3.30) for production functions are fulfilled for Y_{L1} and Y_{L11} by α and violated by β; for γ it depends on the sizes of the factors and technology parameters. But there is no partial or possible violation of the law of diminishing returns for L and E, because *beyond a certain point,* the prohibition of negative output elasticities, i.e. the relations (3.11), prevents L from growing beyond the boundary of the factor space, that only is accessible to Y_{L1} and Y_{L11} because of the restriction $\gamma \geq 0$. Similarly, $Y\beta/L$, the marginal productivity of labor, which decreases with increasing E for Y_{L1} and Y_{L11}, cannot become negative because the condition $\beta \geq 0$ excludes the region of factor space where this would occur. This is matched by the fact that after E grows to the size $E_0 c K_m / K_0$ in the state of maximum automation, a further increase in energy input simply cannot be absorbed by the fully utilized capital stock K_m (Similarly, the time evolution in the direction of increasing automation is represented by the fact that for the LinEx function the $L - E$ isoquant is concave [87, 89]; the $L - K$ and $E - K$ isoquants remain convex).

The simplest production functions, which depend linearly on L or K (as in Eq. (3.23) or Eq. (3.29)) and exponentially on factor quotients, are calculated in Appendix A.4. But because empirical values of Y and K, L, E are not available for the stages of early industrialization and total digitalization, quantitative calculations of economic growth and output elasticities are not possible for these stages.

Examples of higher LinEx functions in which the output elasticities do not simply depend on the first powers of the factor quotients are given in [2, 89] and [115]. Test calculations did not yield improved reproductions of observed economic growth compared to Y_{L1} and Y_{L11} that would have justified the higher mathematical effort.

3.3.6 The Power of Energy and the Weakness of Labour

Table 3.1 shows that in highly industrialised economies such as the USA and Germany the mean output elasticity of labour, $\overline{\beta}$, is much smaller and that of energy, $\overline{\gamma}$, much larger than the respective shares (about 70% and 5%) of these production factors in the total costs of value creation. Earlier studies, limited to the second half of the twentieth century, find the same for Japan [2, pp. 212–213].

Econometric analyses by Ayres and Warr for the USA and Japan between 1900 and 2005 (with the exception of the period of World War II) come to the same conclusion [120]. These authors use as energy variable in the LinEx function $Y_{L1}(K, L, E; t)$ a quantity called *useful work,* which represents the exergy directed by the capital stock (and at the beginning of the twentieth century also by work animals) onto matter and thus already contains efficiency improvements of the energy conversion plants of the capital stock.

Cointegration analyses for Germany, Japan, and the U.S. [121] also affirm the power of energy and the weakness of labor.

This theoretical finding also fits with the observed economic development in the G7 countries FR Germany, Canada, France, Italia, Japan, United Kingdom, USA. This is illustrated in Tables 3.2 and 3.3 by the changes shown in the contribution of the agriculture, industry and services sectors to value added and by the changes in the number of persons employed in these sectors. The agriculture sector (A) includes agriculture and livestock, forestry and fishing; the industry sector (I) includes production of goods in factories and workshops, construction, mining, electricity, gas and water; the services sector (S) includes trade, and insurance, transport, telecommunications, education, education, administration, medical care, nursing, catering, paid domestic work and other small services.

In pre-industrial societies, prosperity was mainly based on primary agricultural production, which also employed most people. This changed fundamentally with industrialisation and the accompanying rural exodus. Nevertheless, as already mentioned in connection with Table 2.2, in the Federal Republic of Germany in 1950, 25% of the workforce was still employed in agriculture and produced 11% of GDP. The mechanisation of agriculture, which began in the USA between the two world

Table 3.2 Value added (GDP) of G7 countries in agriculture (A), goods-producing industry (I) and services (S); as a percentage of total GDP [122] (Totals do not add up to 100 due to rounding errors)

	A			I			S		
Country	1970	1992	2009	1970	1992	2009	1970	1992	2009
FRG[a]	3.4	1.2	0.9	51.7	39.6	27.1	44.9	59.2	72.0
Canada	4.2	2.7	2	36.3	31.5	28.4	59.4	65.9	69.6
France	6.9	2.9	2.1	41.5	29.7	19.0	51.6	67.4	78.9
Italia	8.1	3.2	2.1	42.6	32.3	25.0	49.4	64.6	72.9
Japan	5.9	2.1	1.6	45.1	39.4	23.1	49.1	58.5	75.4
UK	2.8	1.7	1.2	42.5	31.7	23.8	54.6	66.6	75
USA	2.7	1.9	1.2	34.1	28.5	21.9	61.8	68.3	76.9

[a]Until 1990 only the old Federal Republic of Germany before reunification

Table 3.3 Employment structure of the G7 countries as a percentage of all civilian employees in agriculture (A), goods-producing industry (I) services (S) [122]

	A			I			S		
Country	1970	1992	2009	1970	1992	2009	1970	1992	2009
FRG	8.6	3.1	2.4	49.3	38.3	29.7	42.0	58.5	67.8
Canada	7.6	4.4	2	30.9	22.7	22	61.4	73.0	76
France	13.5	5.2	3.8	39.2	28.9	24.3	47.2	65.9	71.8
Italia	20.2	8.2	4.2	39.5	32.2	30.7	40.3	59.6	65.1
Japan	17.4	6.4	4	35.7	34.6	28	46.9	59.0	68
UK	3.2	2.2	1.4	44.7	25.6	18.2	52.0	71.3	80.4
USA	4.5	2.9	0.6	34.4	24.6	22.6	61.1	72.5	76.8

wars and spread to Europe after the Second World War, has since then increasingly replaced human muscle labour with the work of agricultural machinery powered by cheap diesel fuel and electric milking machines and robots. Tables 3.2 and 3.3 show that in the G7 countries between 1970 and 2009, agriculture's contribution to GDP fell to 2% or less, and the number of people employed fell to 4% of all people employed, or less. What has taken place in agriculture is also taking place in industry with a time lag: energy-driven machines are increasingly replacing routine human labor. The people no longer needed in industry are (still) finding work in the expanding service sector.

Because of the high cost share of labour of around 70%, the products of the labour-intensive services sector are expensive and their share of GDP is the largest. However, this does not mean that physical production in agriculture and industry has decreased, on the contrary. Despite declining employment, the quantities of food and manufactured goods produced remain constant, or even increase. Productivity is growing, thanks to rationalization measures of both a technical nature—driven by increasing automation, cheap, production-powerful energy and innovation—and an organizational nature. As a result, most agricultural and industrial products have become cheaper and cheaper, so that their contribution to GDP has decreased.

The problem, which already occurs during phases of weak economic growth, namely that with growing automation production-powerful, cheap energy-capital combinations displace production-weak, expensive human routine work from more and more production processes, is addressed in Sect. 6.1.

3.4 Technological Constraints

Figures 3.2 and 3.3 show that after the oil price shocks, energy use in Germany, Japan and the USA grew more slowly than before. On the one hand, the energy efficiency of production plants was improved in response to the rise in energy prices. In an industrial economy, however, this is only possible until the limits to technical efficiency improvements set by the first two laws of thermodynamics are reached. In other words, technical processes are associated with certain minimum energy requirements. Energy savings usually require investments whose profitability depends on the energy price. Examples of this from the field of rational energy use are given in Sect. 2.3.3. On the other hand, energy-intensive enterprises have been and are being relocated abroad and the service sectors expanded.

Overall, the decoupling of energy and economic growth that was occasionally invoked even long after the first two oil price shocks did not take place *globally*. According to Fig. 2.1, world energy consumption more than doubled between 1970 and 2011. And subsequently, it rose from 18.5 MtC to 19.6 MtC (=13.7 Mtoe) in 2014.

Since the oil price shocks, the price of oil has acquired an economic significance on a par with that of wage settlements with trade unions and interest rate changes by central banks.

Why is this so? Why did the fluctuations in the oil price shown in Fig. 3.1 have such noticeable economic and political consequences? How are the high output elasticities of energy shown in Table 3.1 compatible with the low shares of energy costs in total factor costs?

The resolution of the diamond-water paradox by the "marginal revolution" mentioned in Sect. 3.1 also helps here:

- Water is essential and diamonds are irrelevant to the existence of life on earth. Yet abundant water is cheap, while diamonds are expensive according to their scarcity. This is so because the marginal utility of consuming the last unit of water is so small that when the price of water rises, one may as well do without the last unit, e.g., by shortening shower times, using the toilet more sparingly, and watering the lawn less frequently, until the scarcity of water in our latitudes with good water supply infrastructure is remedied. Similarly, when energy prices rise, the loss of utility to *consumers* can be small if people hold off on buying energy-intensive goods and services that you do not necessarily need for daily living, and the decline in demand for those products does not lead to layoffs of workers in the industries that make them. Who needs the latest smartphone every few years, the most powerful SUV, the most fashionable outerwear, the most stylish home furnishings?—However, if such products are status symbols for people of certain age, occupation or income groups, then their purchase has such a high benefit in terms of status gain for these people that only a drastic increase in the price of the energy required for their production and use would dampen demand.

- The renunciation of the last energy units in the case of cost increases has a significantly different effect on *producers*: It immediately makes itself felt in a noticeable loss of utility, because machines with less energy also work and produce less. In the case of a permanent reduction in plant utilisation, companies can quickly become loss-making, consequently shutting down production capacities and reducing employment, which in turn is associated with a loss of income and other negative economic effects.
- If water is so important to life that a diamond prospector near thirst in the Namib Desert would give up his collected diamonds for a well-filled water bag—why do not we promote life in our latitudes as well by concentrating our investments on the ever more abundant provision of the best drinking water? On its own, this question is silly—death by drowning painfully demonstrates time and again the biological limitation on water intake by humans (and all land creatures). But it is helpful in the context of asking why, given the high production power of cheap energy and the low production power of expensive labor, so much energy is not injected into the production system and so much labor is eliminated from it until output elasticities and factor cost shares are equal for both factors. The obvious answer is: Just as man cannot become more and more efficient by taking in more and more water, so an industrial production system cannot take in more energy at any given time than the energy conversion plants can cope with on the basis of their technical design. To put it bluntly: if you put too much energy into a machine, it will break down. In technical-economic terms: the capital stock can only absorb energy up to the maximum possible degree of utilisation $\eta = 1$. In addition, no more work can be replaced by energy-capital combinations than the degree of automation $\rho \leq 1$ present at the time t and the targeted degree of utilisation $\eta \leq 1$ allow.

3.4.1 In the Blind Spot of Textbook Economics

In the following, we formulate the restriction of energy consumption analogous to water consumption mathematically—on the one hand, in order to explain why the neoclassical economist regards the equality of output elasticities and factor cost shares as a fundamental prerequisite of growth theory, and on the other hand, in order to show that in the blind spot of the observer's eye lie precisely the technological constraints, because of which this prerequisite has been wrong so far.

The degree of capacity utilization η and the degree of automation ρ of the capital stock depend in the following way on the production factors [2, 102, 124]:

$$\eta(K,L,E) = \eta_0^* \left(\frac{L}{K}\right)^{\lambda} \left(\frac{E}{K}\right)^{\nu}, \quad \rho(K,L,E) = \eta \frac{K}{K_m(t)} \tag{3.46}$$

The parameters η_0^*, λ, and ν can be determined from empirical data on capacity utilisation. An example is given in Sect. 3.4.2. $K_m(t)$ is the maximum automated capital stock introduced in Sect. 3.3.5, which at time t could produce a given output Y_t with energy E_m proportional to K_m and a very small number of employees. Furthermore, in a technological state given at time t, the degree of automation of the capital stock for $\eta = 1$ has an upper bound $\rho_T(t)$. This depends on the mass and volume of energy conversion equipment and information processors that the production system must accommodate in the production space if Y_t is to be produced. The outermost limit of automation is reached with $K = K_m(t)$ and $\eta = 1$ at $\rho = 1$. Consequently, the technological constraints that exist for the combinations of capital, labour and energy, are:

$$\eta(K,L,E) \leq 1, \qquad \rho(K,L,E) \leq \rho_T(t) \leq 1. \tag{3.47}$$

These constraints must not be overlooked if the state that a market economy is assumed to occupy at any time is understood as an equilibrium state resulting from the maximization of profit or time-integrated social utility (overall welfare).[14] The maximization of profit is outlined below. Optimizing overall welfare [2, 102, 124] yields almost the same results.

To simplify the notation, we identify K, L, E with the components X_1, X_2, X_3 of the vector

$$\mathbf{X} = (X_1, X_2, X_3) \equiv (K, L, E). \tag{3.48}$$

In this notation and with the help of the slack variables $\mathbf{X}_\eta$ and $\mathbf{X}_\rho$, the constraints (3.47) are put into equation form:

$$f_\eta(\mathbf{X}, t) = 0, \quad f_\rho(\mathbf{X}, t) = 0. \tag{3.49}$$

The slack variables for labour, energy and capital are L_η, E_η and K_ρ. They define the region in factor space within which the factors can be varied independently at time t. Substituting them into Eq. (3.46), we obtain the explicit form of Eq. (3.49) to be

$$f_\eta(\mathbf{X}, t) \equiv \eta_0^* \left(\frac{L + L_\eta}{K}\right)^{\lambda} \left(\frac{E + E_\eta}{K}\right)^{\nu} - 1 = 0, \tag{3.50}$$

[14]The behavioral assumptions that firms maximize profit and individuals maximize utility originate from microeconomics. These optimisation assumptions are also transferred to macroeconomics [103, 125].

$$f_\rho(\mathbf{X},t) \equiv \frac{K+K_\rho}{K_m(t)} - \rho_T(t) = 0 \ . \tag{3.51}$$

3.4.2 Profit Optimisation

The factors of production (X_1, X_2, X_3) may have the prices per unit factor set by the market, $(p_1, p_2, p_3) \equiv \mathbf{p}$. Then the sum of factor costs is $\mathbf{p(t)}\cdot\mathbf{X(t)} = \sum_{i=1}^{3} p_i(t)X_i(t)$. Economics defines *equilibrium* as a state in which the profit

$$G(\mathbf{X},\mathbf{p},t) \equiv Y(\mathbf{X},t) - \mathbf{p}\cdot\mathbf{X} \tag{3.52}$$

is maximal.[15] The necessary condition for the maximum of profit subject to the technological constraints (3.50) and (3.51) is:

$$\vec{\nabla}\left[Y(\mathbf{X};t) - \sum_{i=1}^{3} p_i(t)X_i(t) + \mu_\eta f_\eta(\mathbf{X},t) + \mu_\rho f_\rho(\mathbf{X},t)\right] = 0. \tag{3.53}$$

Here $\vec{\nabla} \equiv (\partial/\partial X_1, \partial/\partial X_2, \partial/\partial X_3)$ is the gradient in factor space; μ_η and μ_ρ are Lagrange multipliers (The *sufficient* condition for the profit maximum contains a sum of second derivatives. It is assumed that the extremal value of the gain at finite X_i is the maximum).

From Eq. (3.53) follow the three equilibrium conditions

$$\frac{\partial Y}{\partial X_i} - p_i + \mu_\eta \frac{\partial f_\eta(\mathbf{X},t)}{\partial X_i} + \mu_\rho \frac{\partial f_\rho(\mathbf{X},t)}{\partial X_i} = 0, \quad i = 1,2,3. \tag{3.54}$$

Multiplying (3.54) by X_i/Y and writing the output elasticities defined in Eq. (3.4) α, β, γ as

$$\epsilon_i \equiv \frac{X_i}{Y}\frac{\partial Y}{\partial X_i}, \quad i = 1,2,3, \tag{3.55}$$

the equilibrium conditions become

$$\epsilon_i \equiv \frac{X_i}{Y}\frac{\partial Y}{\partial X_i} = \frac{X_i}{Y}\left[p_i - \mu_\eta \frac{\partial f_\eta}{\partial X_i} - \mu_\rho \frac{\partial f_\rho}{\partial X_i}\right], \quad i = 1,2,3. \tag{3.56}$$

You can put them in the form

[15] The profit-maximizing equilibrium state of economics corresponds to the equilibrium state of maximum entropy of a closed system in thermodynamics.

$$\epsilon_i = \frac{X_i[p_i + s_i]}{\sum_{i=1}^{3} X_i[p_i + s_i]}, \quad i = 1, 2, 3, \tag{3.57}$$

by summing the left and right sides of Eq. (3.56) via $i = 1, 2, 3$, noting that because of (3.10) $\sum_{i=1}^{3} \epsilon_i = 1$; in the resulting equation, Y must then be expressed by the other terms.

The s_i are

$$s_i \equiv -\mu_\eta \frac{\partial f_\eta}{\partial X_i} - \mu_\rho \frac{\partial f_\rho}{\partial X_i}. \tag{3.58}$$

They represent the technological constraints in monetary terms. As a result of our *non-linear* optimization, we call them *generalized* shadow prices—"*generalized*" to avoid confusion with the term "shadow prices", which refers to the corresponding expressions obtained in linear optimization.

The Lagrange multipliers μ_η and μ_ρ depend on the factor prices p_i, the production function Y and the derivatives of $f_\eta(\mathbf{X}, t)$ and $f_\rho(\mathbf{X}, t)$; the details are in [2, 102]. For its part, Y can only be computed if the output elasticities ε_i are known. The equilibrium conditions (3.57) could only provide output elasticities if the optimum was not at the margin but *within* a region of *K, L, E space* that would be accessible to the system despite the technological constraints, so that it would not even feel these constraints locally. This was not and is not the case with the prices of capital, labour and energy as they had formed and are forming in the markets so far. Figure 3.7 shows an example of how the constraint on capacity utilization has prevented an economic system from sliding down the cost mountain to where profit would be maximized.

In a world without technological constraints, *all of K, L, E space* would be accessible to the economic system. The optimization would take place without the Lagrange multipliers μ_η and μ_ρ, the generalized shadow prices would be zero, and the equilibrium conditions (3.57) would be the cost-share theorem: the right-hand side of (3.57) would have in the numerator only the costs $p_i X_i$ of the factor X_i, the denominator would be the sum over all factor costs, and the quotient, which is equal to the output elasticity ε_i, would represent the share of the costs of the factor X_i in the total factor costs. But there is no such thing as an economy without technological constraints.

Sidenote

In the absence of technological constraints, there would also exist the duality between factors of production and factor prices, according to which all necessary economic information should already be contained in the factor prices. This duality, which is often used in orthodox economic analyses, follows from the Legendre transformation, which in turn results from maximizing the profit function $G(\mathbf{X}, \mathbf{p})$ *without* regard to technological constraints. Indeed, Eq. (3.54) would then hold with

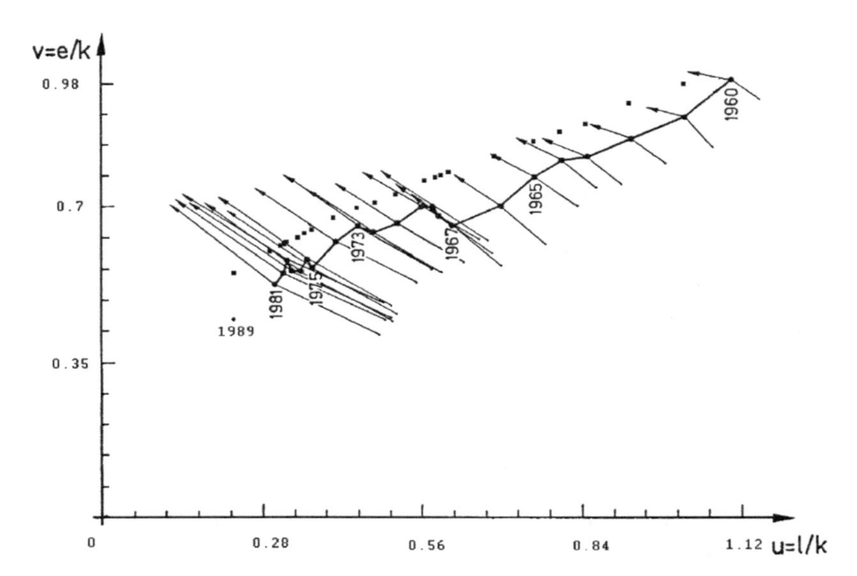

Fig. 3.7 The path of German industry ("goods-producing industry") in the cost mountain between 1960 and 1981 (solid line), direction and strength of the cost differential (arrows/dashes) and the capacity wall of maximum utilization $\eta = 1$ (full squares), projected onto the plane spanned by $u = l/k \equiv \frac{L/L_0}{K/K_0}$ and $v = e/k \equiv \frac{E/E_0}{K/K_0}$ [124]

$\mu_\eta = 0 = \mu_\rho$, and from this one could determine (analytically or numerically) the factor inputs $X_{1M}(\mathbf{p})$, $X_{2M}(\mathbf{p})$, $X_{3M}(\mathbf{p})$ at the profit maximum. With $\mathbf{X}_M(\mathbf{p})$ the profit function (3.52) would become the price function

$$G(\mathbf{X}_M(\mathbf{p}), \mathbf{p}) = Y(\mathbf{X}_M(\mathbf{p})) - \mathbf{p}\cdot\mathbf{X}_M(\mathbf{p}) \equiv g(\mathbf{p}) \tag{3.59}$$

All essential information on production would be contained in the price function $g(\mathbf{p})$. This is the Legendre transform of the production function $Y(\mathbf{X})$ (Here again there is a formal analogy with physics, namely with the Hamilton function of classical mechanics, which is the Legendre transform of the Lagrangian function, or with enthalpy and free energy of thermodynamics, which are Legendre transforms of internal energy). But because of technological constraints and generalized shadow prices, the cost-share theorem and Eq. (3.59) are special cases that do not include production systems with the factor price ratios given so far. Rather, for the economic systems we have known so far, the shadow prices s_i in Eq. (3.57) are nonzero.

3.4.3 Downhill, Along the Rampart

The macroeconomic production function (3.1) includes the idea that at any time t the sum of all entrepreneurial decisions can be understood as the decision of a "representative entrepreneur" about the quantities of capital $K(t)$, labour $L(t)$ and energy

$E(t)$ to be used. Does this representative entrepreneur act rationally in the economic sense? In other words: Do real-existing economic systems really develop along a trajectory of equilibrium states defined by profit maximization? As far as this question only concerns factor input ratios, it can also be: Does the representative entrepreneur always seek the cost minimum?

We consider a cost mountain rising above the plain spanned by the factor ratios of labor to capital, $\frac{L/L_0}{K/K_0} \equiv u$, and energy to capital, $\frac{E/E_0}{K/K_0} \equiv v$. Changes in factor prices correspond to snowfall, which can also trigger avalanches, and the resulting changes in the snowy topography of the slope on which the "sled of the economy" is traveling. The representative entrepreneur steering the sled must be mindful of the capacity wall created by the technological constraint $\eta \leq 1$ that restricts its descent. In Fig. 3.7 the cost mountain with its gradient lines and the capacity wall as well as the path of the economic system "FRG Industry" is projected onto the $u - v$-plane. Inflation-adjusted, aggregated factor prices were only available for this system and the period from 1960 to 1981 from a research project [88].

The computation of the cost gradients and the rampart of maximum capacity utilization, $\eta = 1$, is accessible via [124]. It will not be repeated here. We restrict ourselves to explanations of Fig. 3.7 and conclusions drawn from it.

- The direction of the arrows above the path representing the projection of the movement of "FRG Industry" in the cost mountain onto the $u - v$-plane indicates the direction of the cost gradient calculated with the LinEx function Y_{L1}, Eq. (3.42). The arrow lengths correspond to the strengths of the cost gradients. Scaling is to 1970 output.
- The lines without arrowheads below the path were calculated using the energy-dependent Cobb-Douglas function (3.31), whose output elasticities are roughly equal to the output elasticities of the LinEx function averaged over the period 1960 to 1981. The cost differential differs only slightly from that calculated with LinEx.
- The path does not follow the cost differential in general, but only temporarily during the cyclical upswings of 1967–1970, 1972–1973, and between 1975 and 1979. These movements of the system in the direction of the capacity wall occur because after periods of recession with poorly utilized production facilities, increases in energy use can quickly offset previous losses.
- The isolated point for the year 1989 indicates that until German reunification in 1990, the path maintains its general direction of sharply decreasing ratios of labor to capital, u, and moderately decreasing ratios of energy to capital, v. The compression of the annual points with decreasing u *is* due to the increase in the capital stock and decrease in hours worked shown in Fig. 3.2.
- The path is more or less parallel to the capacity wall, except for a clear shift towards the slope lines at the end of the 1960s, when strong growth in industrial production at full employment forced cost-reducing automation and led to growing energy/capital ratio v.

- Over time, the gradients of the cost differential become stronger. Growing automation increases the cost pressure. According to "Moore's Law", the density of transistors on a microchip doubled every 18 months in the last four decades of the twentieth century. Automation could not have progressed any faster. Faster cost reduction along the lines of the cost differential by massively increasing the use of cheap energy has been inhibited by the capacity wall.

The representative entrepreneur who steers the "economic sledge FRG Industry" as shown in Fig. 3.7 always sees the capacity wall. He steers his vehicle parallel to this wall without touching it. This leaves him room to manoeuvre. For it is advantageous not to scrape along the capacity wall, but to be able to choose capital, labour and energy independently of each other in such a way that the entrepreneur can adjust them as best as possible to the fluctuating demand for goods and services. This is why the economically optimal capacity utilization is less than 100% from the outset. If an economic system operates at, say, 90% capacity utilisation, the cost of adjusting to increasing or decreasing capacity demand is less than in a state of 100% capacity utilisation. Overtime is expensive, as is machine downtime and underemployed workers. In addition, machines require maintenance. If they are constantly running at full capacity, maintenance is difficult to accommodate.
In this sense, the technological constraint "upper bound on capital stock utilization" operates even before the rampart it erects in the cost mountain is reached. We call it *virtually* binding. Like a slightly skidding racing sled in the ice chute, the system slides downhill parallel to the capacity wall.

On the development path of "FRG Industry" shown in Fig. 3.7, expensive routine work is replaced by cheaper energy-capital combinations. On this path, the representative entrepreneur satisfies the growing demand for goods and services, e.g. from the field of information technology, for the value-added production of which he has to expend less energy than, for example, for digging deep excavation pits and melting steel girders. Moreover, he increasingly uses imported intermediate products whose energy content, i.e. the energy used to produce them, is not included in his energy balances. On the other hand, he has to respect legal and social obligations that prevent the dismissal of workers, even if quick cost reduction suggests it. He may also appreciate the long-term benefits of well-trained, loyal workers and employees. These additional, "soft" constraints complement the technological constraints.

Overall, (moderate) cost minimization seems to have been practiced by economic actors in German industry during the Cold War. It would be desirable to examine the behavioural assumption of "cost minimization" with data also for the period after the fall of the Berlin Wall, when the production system on the territory of the defunct GDR was adapted to that of the FRG. In this context, Jürgen Grahl's [126] suggestion could also be taken into account to model the (only virtually binding) technological constraint and the soft constraints with respect to capital stock utilization by a penalty function in profit, which drives up costs when approaching the capacity wall.

Post-growth Economics 4

4.1 Sustainability and Growth

Measured in terms of the overall balance of their effects, all previous sustainability efforts have failed, regardless of whether they were technological, political, educational or communicative measures. Apart from insignificant exceptions, no ecologically relevant field of action can be found in which the sum of long-known and new damaging activities has not permanently increased. The historical enterprise of modernity has drifted off course, cultivating promises of freedom that resemble a bad check, because the calculation was made without the Second Law of Thermodynamics. What is celebrated as progress in civilization threatens to end in a dead end. In fact, the increase of symbols of wealth—even far beyond the satisfaction of basic needs—is given priority over the preservation of the natural foundations of life.

Even social niches in which ecologically compatible models of life were located in the late 1970s and early 1980s have more or less succumbed to the lure of a comfortable, geographically delimited existence. Their once pioneering impulse has long since sunk into a tide of material armament, digitalization, disposable littering, and exploding demand for long-distance travel. The number of educated, politically progressive people whose global radius of action corresponds to an individual CO_2 footprint that exceeds anything previously thought possible is growing by leaps and bounds.

What seems urgently needed is not only a reappraisal of the current model of prosperity, especially its destructive side-effects, but also an economic blueprint for the future that is consistently oriented towards sustainable development. In the present chapter, some of the origins and consequences of the industrial system, which in the view of many experts cannot be stabilized without economic growth, will first be examined from the perspective of post-growth economics. It is aimed at avoiding the catastrophe mentioned in Sect. 1.1.3 and feared by the Nobel Prize-winning economist Robert M. Solow.

R. Kümmel et al., *Energy, entropy, creativity*,
https://doi.org/10.1007/978-3-662-65778-2_4

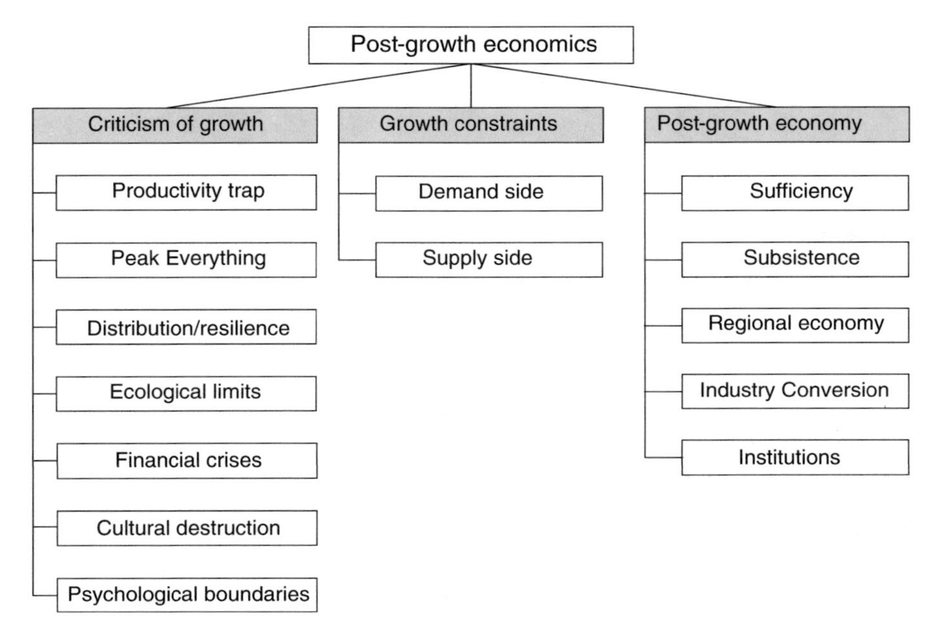

Fig. 4.1 Subject matter of post-growth economics

As a teaching and research programme, post-growth economics focuses on three fundamental questions (Fig. 4.1):

1. What rationales suggest that further growth in global value creation within our planet's biosphere cannot be an option for the twenty-first century?
2. What are the reasons why a growth imperative is seen for modern, globalized, industrial economies?
3. What are the contours of an economy beyond further growth (post-growth economy)?

Post-growth economics is a heterodox sub-discipline of economic sustainability research that is oriented towards the principles of thermodynamics. It denies the optimism of progress that material scope for action can be increased in the finite system of the Earth. Instead, it is based on a material zero-sum logic: more material freedom is inevitably bought with a loss of usable resources and an increase in ecological damage.

This premise blatantly and momentously contradicts contemporary constructions of modernity that implicitly assume that economic surpluses can be created and distributed out of material nothingness through increased efficiency, knowledge and creativity. Post-growth economics establishes the insight that a socially just state of affairs can only be achieved if it is accepted that the available mass of distribution is limited. Because what results from industrial specialization cannot be achieved without ecological plunder and therefore requires limitation. The term "plunder" is used here and in the following as a synonym for an unsustainable exploitation of natural resources (living beings, materials, energy sources and emission absorption capacity of the biosphere).

Since it is impossible, not only empirically evident but also theoretically easy to demonstrate, to decouple permanent growth of material prosperity from ecological damage (which will be discussed in the following sections), the question arises: What material freedoms may a single person claim without living ecologically and thus at the same time socially beyond his means? Those who take too much for themselves reduce the scope for action of people living elsewhere and/or in the future. Social progress, which is based on achieving a surplus and distribute it justly, which in a fairer world should not have been created at all, leads to a contradiction. The resulting zero-sum logic translates the consequences of thermodynamics right into the realms of distribution and social policy. This highlights the fact that modern conceptions of development and progress, still based on unrestrained expansion, are dysfunctional in the long run. What is declared as surplus or distribution mass in this context is pure substance depletion on our finite earth.

Insofar as this applies to contemporary industrial and consumer societies, the need arises to question established claims to material freedoms that cannot be justified—i.e. cannot be transferred globally and intertemporally. The necessary reductions cannot be delegated to anyone or anything; they can only be implemented at the individual level, regardless of whether they are imposed by a political authority or result from autonomous cultural change. A corresponding economic concept, which is called "post-growth economy" [127], has first of all to master the challenge of organizing sufficiency in a socially acceptable way. Consequently, patterns of supply and lifestyles must be developed that are consistent with a dignified existence within an economy which is significantly smaller than current consumer societies and is no longer growing.

4.2 Overcoming Zero-Sum Logic as a Modern Myth

It is not by chance that economic growth and the continual increase in technical possibilities have become characteristic of modern ideas of development. The increase of human scope for development and material wealth has always been seen as a means to alleviate distribution conflicts and to create peace. Erhard's famous statement [128] that it is better to let the economic "cake" grow instead of fighting over the distribution of a given surplus has become the symbol of a political culture according to which social cohesion can be bought with economic growth. If the mass of distribution grows, interest groups can be served without having to diminish the absolute status quo of other groups. For democratically legitimate decision-makers, it therefore causes lower political opportunity costs to reinterpret distributional conflicts as growth targets, rather than solving them by redistributing what already exists.

Origins for this paradigm can be found in a transformation of occidental philosophy, which was dominated by a world view called "limited goods conception" until the Copernican turn. With reference to the Greek conception of a cosmos as a finite whole, Jaeger [129] speaks of a "legal community of things". Analogies to the game-theoretical concept of the zero-sum game become clear, which Haesler [130] characterizes as follows:

- all beings, things and resources are finite;
- nothing can come into being in this cosmos without something else perishing (as a result);
- nothing can arise from itself, no creatio ex nihilo;
- nothing can perish without replacement;
- the change of things is constant transformation and compensation;
- the sum of gains and losses is zero; one man's gain is another man's loss; there are no transactions in which the advantage can be on both sides;
- every being, thing and resource has a certain measure (metron) which determines its place (topos) in the cosmos;
- if this order is violated, then this "injustice" is cured by the "decree of time"; this is the goddess Diké's "clock" that strikes back when it pleases.

The overcoming of this zero-sum game can not only be traced as a leitmotif of the history of economic dogma, but also forms the epitome of scientific-rational conceptions of progress in modernity, which Francis Bacon already associated with a comprehensive mastery of nature. The notion of instrumental reason derived from this has lost none of its relevance to this day: Earthly resources are to be converted into means from which an unlimited spiral of increase in human well-being is to be fed. The term "substantial progress" coined by Simmel [131] ties in with this: If it were possible to increase the stock of what exists, the "human tragedy of competition" could be alleviated, namely in the form of a "diversion of the struggle against man into the struggle against nature" [130, p. 65]. "To the extent that one draws further substances and forces from the still unoccupied supply of nature into human enjoyment, those already occupied are relieved of the competition for them" [131, p. 305]. Relief, however, does not mean overcoming, since the aforementioned "substances and forces of nature" are, as Simmel quite concedes, finite. Therefore, he sees the real key to "building up a world that can be appropriated without dispute and mutual displacement, to values whose acquisition and enjoyment on the part of one does not exclude the other, but opens the way to the same a thousand times over for the other" [131, p. 306], in a broader vision, namely that of "functional progress". This refers to the processes of exchange made possible by the introduction of the modern monetary economy. The progressive monetarization of all areas of life can be seen as an act of liberation of the individual—the peasant henceforth no longer produces (exclusively) for the lord of the manor, but for the market—and as a distancing of the individual from the previously binding social contexts.

The increasing "materialization of interaction" through money-mediated exchange results in its expansion and progressive interconnection. The medium of exchange money ensures that all things pass into "the more fertile hand" in order to "release a maximum of the value latent in them". This heralds a profound structural change. It leads to the emergence of new motivational structures, efficient forms of division of labor, and a growth dynamic designed to turn the negative into a positive-sum game. "Assuming that the world were really 'given away' and that all action really consisted in a mere shifting back and forth within an objectively unchanging quantum of value, the form of exchange nevertheless brought about, as it were, an intercellular growth of values" [131, p. 307]. This logic is reminiscent of a mixture of perpetual motion and eternal cornucopia; moreover, it arouses associations with the special case of public goods. "Thus one might suppose that Simmel's world of goods is an immense theatrical hall in which the enjoyment of one not only does not exclude that of the other, but conditions it" [130, p. 72].

This epochal undertaking is believed to contribute to the civilization of mankind, because in the course of a comprehensive and visibly differentiated division of labor, everyone could participate. Peacefully united in busy plundering, one crow does not peck out another's eye. The interconnectedness and division of labour processes that go hand in hand with this embed everything social in economic relationships. According to the modern reading, this gives rise to peacemaking ties. For those who maintain complex trade relations for mutual benefit (usually) do not wage wars, so the hope goes.

The European integration process also follows precisely this logic. The penetration and exploitation of an enlarged economic area, it is said in every political Sunday speech, serves peace and social cohesion. This requires a uniform currency, i.e. the abandonment of national sovereignty, which would bring people closer together and could spur an equalisation of living conditions. This intention seems to justify many political and economic means, some of which turn out to be ecologically ruinous. Thus, the face of the much-invoked progressive Europe reveals not only a common identity and political harmonization, but sprawling industrial plants, agro-economic monocultures, resource intensive infrastructures, unconditional digitalization, growing construction activities, and fossil-fuel mobility.

This leads to two conclusions: (1) In order to bring human civilizations into a state of peace with the help of permanently increasing distribution masses, an even more intensive war against the ecosphere is obviously necessary. (2) Modern growth strategies, which presume to foster human well-being, are based on disregarding the zero-sum logic of thermodynamics, according to which every production of a good also produces negative outputs, such as emissions.

4.3 "Green" Growth: Continued Plunder by Other Means?

Although the present patterns of increase [132, 133] has long been confronted with physical limits, it is constantly revived anew. In this process, only its material manifestations are differentiated. Paths of expansion of an older type are supplemented, stabilized or strengthened by those of a newer character. Despite a dramatic shortage of ecological source and sink functions, the growth regime as such is not called into question, but only steered in other directions.

In order to counter resource bottlenecks, attempts are made, true to the substantial progress logic of Simmel, to use basic technical innovations to (a) open up new deposits for resources that have been used up to now (e.g. more intensive extraction measures for oil reserves in deep ocean regions or under melting polar ice), (b) find substitutes for dwindling resources (e.g. extraction of oil or tar sands; fracking) or (c) redefine previously "useless natural assets" into exploitable resources (e.g. conversion of landscape conservation areas into wind farms) or (d) redefining previously "useless" natural assets as exploitable resources (e.g. converting landscape conservation areas into wind farms, using agricultural land as cultivation areas for energy crops).

Simmel also seems to have anticipated the reactions to the even more glaring problem of overloaded sink functions, especially as a result of the consumption of fossil fuels. Analogous to his functional concept of progress, dematerialized, i.e. "intercellular growth" has been propagated since the mid-1970s. In the sense of ecological modernisation, the aim is to decouple increases in gross domestic product (GDP) from ecological damage by means of technical progress. This is to be made possible by three categories of measures. Firstly, efficiency measures aimed at minimising the resource input per unit of value added [134, 135], secondly, closed resource cycles to enable an industrial metabolism that is ecologically harmless [136, 137], and thirdly, the use of renewable energy sources to create decarbonised production and transport systems, i.e. a "solar world economy" [138].

The relief of the ecosphere is thus delegated to a technological development in order to be able to leave mobility- and consumption-based forms of individual development largely untouched. The fate of humanity is thus dangling from a vague promise of innovations. It is left to a technical progress that has not yet taken place or become effective and whose future occurrence is unprovable, i.e. can only be hoped for. Furthermore, it cannot be known in advance what unintended side-effects innovation-driven progress will reveal and whether these will possibly counteract everything that it promises to contribute to solving problems even under theoretically optimal conditions [139].

Compared to this, the late medieval paradigm of nature control à la Bacon and Descartes at least has the mitigating circumstance that ecological risks—especially the consequences of thermodynamics—could not be known. The optimism of progress was limited to being able to occupy a seemingly inexhaustible supply of nature and transform it into goods, which has so far succeeded to an increasing extent and now leads to the problems described in Chap. 1 because of the law of entropy. In contrast, the confidence in progress of green growth is not content with merely continuing to increase the materialized symbols of freedom and well-being, but proclaims on top of that that it can accomplish this in an ecologically harmless way, which amounts to a double effort of faith. Behind what pretends to be ecologically enlightened is thus only an even more abstruse technological mania for feasibility. This is exemplified by climate change, of all things the most pressing of all sustainability deficits.

Despite a flood of climate protection innovations, ecological damage in the energy sector—both greenhouse emissions and the destruction of nature and landscapes—has steadily increased. The proof that it has ever been possible in practice to solve an ecological problem by means of technical solutions with a holistic consideration of all spatially and temporally distant but environmentally relevant (side) effects is still lacking.

4.4 How Advanced Can Technical Progress Be?

Let us imagine the following experimental setup: A tinkerer who already holds several patents is looking for a new challenge. He takes two peas in each hand. He brings both hands together to form them into a hollow body, shakes the objects inside, closes his eyes, opens his hands and lets the contents fall down. Now he opens his eyes, searches the ground, and hopes to find five peas. With frequent repetition of the experiment and sufficiently strenuous shaking, he believes, Adam Riese's laws would eventually be empirically overcome. No less grotesque is the adherence to the decoupling paradigm, since it resembles a stoic race against the laws of thermodynamics. Once again, this is a manifestation of the epochal delusion—actually glorified as modern and progressive—that additional material scope for action can be created solely from applied knowledge or human creativity, bypassing the laws of nature.

Against this background, Neirynck [140] has tried to sketch an evolution of technology. Some of his remarks raise the question whether technical progress in an industrial economic system limited to the biosphere of the earth can be possible at all in the long run, at least if it is understood as transforming a current state into a "better" one, namely in such a way that the latter reveals more human freedoms of action, wealth, convenience or solutions to problems than the previous situation.

For Neirynck, technical progress means that physical facts are transformed into a changed or new order. This is done with the help of human knowledge. The new physical order serves to create additional material possibilities, whether through a hand axe hammered out of a basalt column, the production of cars, the development of arable land, or the construction of a coal-fired power station or a solar plant. But none of these selectively and temporarily created physical orders, from which additional possibilities for action arise, come solely from knowledge or creativity: for nothing can arise without the use of the production factor energy, and inevitably connected with its transformation into value-creating physical work is entropy production in the form of particle and heat emissions.

If their effects are not sufficiently neutralized by heat radiation into space and the energy of sunlight, they lead to ecological damage. Insofar as these are often only perceptible as spatially, temporally, systemically or materially displaced phenomena, it is easy to suggest that they are only sporadic, not necessarily occurring side-effects that can in principle be avoided or eliminated. Insofar as it is assumed that the effort required for this does not overcompensate for the benefit of the causal economic activity, a surplus remains that seems to justify the continued assumption of progress.

Economic ideas of progress are usually based on being able to generate increased value creation or additional human benefits from a given input of resources. Such an improvement in the means-purpose relationship, which is understood as increased productivity or efficiency, can also consist in reducing the use of resources to achieve a certain result. Even without a change in technology, increases in efficiency are possible in principle, for example by redistributing given quantities of goods or resources and directing them to where they are most urgently needed, i.e. where they provide the greatest benefit. Such an increase in allocative efficiency, associated with the name of Pareto, can also be seen as a specification of Simmel's functional progress. Although the increases in efficiency brought about by mere redistribution (as opposed to technical progress) may indeed be immaterial in nature, they reach a material limit quickly.

This is the case when it is no longer possible to improve the situation of an individual through a re-allocation of goods without making another individual worse off. If this zero-sum constellation has occurred, further progress is only possible if the mass of goods available is increased. This is exactly what four industrial revolutions have done. The remarkable increases in production are commonly interpreted as (technical) increases in efficiency. They result from a combination of innovative use of technology, specialisation measures, more intelligent organisational principles and learning processes. In view of the immense ecological degradation and resource depletion that occurred in parallel, one can ask whether the supposed progress is actually based on "real" efficiency or simply more effective plundering.

Efficiency would presuppose that the increases are due to a "smarter" combination or a—technically conditioned—more productive use of the resource endowment used so far. If this were so, it should be possible to achieve steadily increasing material surpluses within a physically closed system of production without the supply of external resources and outsourced damage. Did such surpluses ever exist, and if so, to what extent—if all spatially, materially, temporally or systemically outsourced side-effects were taken into account, which would have to be subtracted from the production result as costs? Schumacher identified in his 1973 classic *Small is Beautiful* [141] as a cardinal error of contemporary economic thinking—whether neoliberal or Marxist—that substance depletion is confused with surplus. But to distribute a presumed surplus "justly", which in a just world should not exist at all, because it is based on irreversible plundering and thus destroys future life perspectives, leads itself ad absurdum [142].

4.5 Externalities and Efficiency

Orthodox economists do address the question of the "true" surplus, but they use the questionable concept of so-called "external effects", which was established by Pigou [143] and Kapp [144], among others. According to this conception, the damages caused by unintended side effects are regarded as a form of market failure, because the affected environmental goods have no price. Consequently, the market mechanism cannot ensure that scarce resources are used efficiently: Firms are not confronted with the full costs of their actions, which are thus not reflected in operational profit maximization.

To correct this systemic flaw, environmental economics recommends "internalizing" negative externalities by imposing the actual (environmental) costs of their activities on polluters, for example through environmental taxes or pollution licences ("cap and trade"). In this way, emitters are motivated to avoid environmental damage and waste of resources, because otherwise costs are incurred that reduce profits. This approach, which not only seems logical but is also widely accepted, does not solve the problem for at least the following reasons.

1. The externalisation strategy suggests that the environmental damage of the modern industrial system is in principle avoidable if only "better" substitutes, technologies, institutions and organisational structures are applied. But what if, for the most ecologically ruinous practices of all (especially air travel), there are no foreseeable alternatives that are more sustainable and at the same time accepted as functionally equivalent from the consumer's point of view? Then only general bans or prohibitively high prices would help, but these are considered reprehensible in terms of distribution policy, or at least unenforceable by democratic means.
2. The theory of externalities does not imply avoiding ecological degradation completely, but only reducing it to an "optimal" level of damage. Since ecologically harmful activities generate economic benefits—otherwise they would not take place—and environmental protection is costly, a trade-off problem results. Accordingly, the reduction of ecological damage would only be rational as long as the costs of environmental protection (including welfare losses) are lower than the benefits of it. Therefore, it is possible to justify theoretically and scientifically what politicians do every day: actions that are ecologically disastrous are justified on the grounds that the resulting benefits and opportunities should be valued more highly than the environmental damage. In case of doubt, it is argued that the competitiveness of the national economy is at stake or that unemployment is imminent. The economic benefit or the socio-political necessity of any intervention in nature can be estimated as high as desired by pure speculation. Consequently, there is no limit to the ecological destruction that can be accepted: the end justifies the means.
3. Trying to remove negative externalities from the growing prosperity amounts to merely optimising the ecological details of an otherwise unquestioned industrial system. This technological optimism prevents a ruthless but overdue reflection on modern industrial and consumer societies: How high would be the remaining, i.e. plunder-free surplus of industrial production after deduction or correction of all environmental damage?

4.6 Economic Efficiency as a Fallacy?

A "pure", i.e. plunder-free efficiency effect was formative over long epochs of humanity before the dawn of the industrial age, i.e. in pre-modern agricultural and craft economies. For example, the transition from self-sufficient shoe production by single individuals to specialized shoemaking operations significantly increased production (not only of shoes) within a village. Learning processes and intra-organisational specialisation reduced waste in leather and shoe sole processing. Moreover, the tools or machines needed had to be purchased only once.

This led to lower average costs, because previously each household needed its own cobbler's tools for its small shoe needs. Improved organisational structures as well as more skilful and concentrated operations were able to increase labour productivity. Thus it became possible to achieve a higher production result with an unchanged input bundle within a given spatial system compared to pure self-sufficiency, or to require fewer scarce resources for the same level of supply; the savings could in turn be used for other goods. Overall, material well-being increased.

But the potentials of this ideal-typical increase in efficiency proved to be extremely limited due to its pre-fossil character, in particular due to

- the limited spatial scope of interorganizational specialization,
- the rapidly reached upper limit of the employees' physical working capacity, and
- the low and limited speed with which resources and units of performance can be moved within the value creation process.

This threefold limitation resulted in a static economy for centuries with a correspondingly modest level of overall production, diversity of options and technical development. Only the spatial, physical and temporal—i.e. triple—dissolution of boundaries with the help of mechanisation, automation, electrification and finally digitalisation, which presupposed the increased availability of fossil energy sources, helped to break out of the limitation and trigger the dynamic of increase that is confused with efficiency. For the effect of the de-limiting arsenal of amplifiers was not so much to increase the productivity of previously used resources by pure creativity quasi out of material nothingness, but rather to pave the way for more effective exploitation and addition of external resources. Instead of an improved goal-means ratio, as would correspond to the definition of economic efficiency, additional or new resources were simply tapped.
How, for example, can the immense increase in agricultural yields (=output) per hectare (=input) seriously be described as an increase in efficiency if the additional energy, chemical, fertiliser, machinery and logistical inputs associated with it are taken into account? Moreover, some of the additional inputs that have increased the yield per hectare so significantly are absurdly themselves the result of other land use in Asia or Latin America (for example, animal feed based on soy products).

Assuming that in the shoemaker example above all pre-fossil specialization potentials within the village were exhausted. How then could further increases in wealth be achieved? First, the cobbler could start to supply a neighbouring town in order to be able to increase sales in such a way that an even lower average cost level is achieved through economies of scale. In turn, bread production in the neighbouring town could be expanded so that the demand for bread in both towns is met by only one producer, who can reduce costs by increasing output accordingly. This leads to a higher degree of specialisation in order to reduce costs, because the sales of both products are now extended beyond the previous radius of action (downstream delimitation).

A further cost reduction could be achieved if the shoe producer specialises further by no longer producing the shoe soles itself, but instead purchasing them from a company that can manufacture more cost-effectively at another location—for example in China or India. Similarly, the bakery could import low-cost "dough" from far away (upstream de-limitation). This again makes the shoes and bread cheaper and consequently increases purchasing power.

But what physical effort does this increasing dissolution of boundaries require? The bakery and the cobbler's workshop are growing to factory size, needing new and larger production facilities, buildings, storage facilities, transport and communication systems, and are permanently expanding the radius of action of their supplier networks and sales channels. The public infrastructure must be adapted to the additional transport, logistics and space requirements. Energy, education and information systems must also be expanded accordingly. This is where industrialization begins. Through it, the above-mentioned limitations of a formerly static economy are broken, whose from now on expansive development spreads to the entire society and expresses itself in material occupation. The dilemma is that the exploitation of "pure", i.e. pre-fossil, efficiency potentials until the "de-limitation point" is reached only allows for a modest and hardly growing prosperity, which could, however, be considered ecologically almost free of plunder. All production levels growing beyond this point are based on a technically intensified and spatially delimited depletion of ecological sources and sinks.

External effects are therefore not a side effect of the remarkable increase in prosperity since the beginning of industrialisation, but its precondition. Similar to a production factor that cannot be substituted, they are indispensable for the growth of goods. In other words: If it were no longer possible to externalize ecological damage, the industrial system would lose its basis. If external effects were consistently avoided, the economic surplus would be little more than the level at the de-limitation point described above.

4.7 Rebound Effects

Nevertheless, the hope of a plunder-free surplus is clung to, which is not only to clearly surpass the pre-industrial level, but to be constantly increased quantitatively and qualitatively. This is understandable: for as untenable as economic efficiency and progress projections are from a thermodynamic point of view, to admit this would be to shake the foundation of faith in modernity. The vision that has guided action for more than a century, according to which peace, justice and comfortable consumer prosperity can be achieved through never-ending economic growth, would be difficult to legitimise. It is not even sufficient to keep GDP constant in order to stabilize a certain material supply level. Binswanger [145] has been able to show that under industrial production conditions already a suspension of the growth dynamic (i.e. no reduction, but merely zero growth) leads to triggering a downward spiral of value creation.

To solve this dilemma, attempts are being made to save the modern vision of prosperity through "green" growth. But the endeavour to decouple economic growth from environmental damage by means of technical innovations has so far failed—especially with regard to climate protection. In sustainability research, the view is increasingly gaining ground that this is not a coincidence, but of systematic origin, namely due to "rebound effects". Two particularly important rebound categories can be identified, the first of which is based on the fact that the elimination of an environmental problem is bought with causing an additional or new environmental problem elsewhere, later or in a different way. For example, if energy-saving light bulbs are used to reduce electricity demand through efficiency, this is countered by the fact that the production and disposal of these light bulbs create a new source of harm simply because of their mercury content. The second category, namely financial rebound effects, is based, for example, on the fact that the savings in electricity costs (to stay with the above example) give rise to additional purchasing power, which in turn can be used for other goods, so that on balance energy consumption does not fall or may even rise.

Both types of rebound, which will be discussed in Sects. 4.7.1 and 4.7.2, do not imply, however, that environmentally friendly technical progress must be ineffective per se. This is particularly true if the relevant innovations are made subject to the proviso that industrial output is to continue to grow. An overview of this topic can be found in Paech [146, 147] and Santarius [148].

Increases in GDP presuppose additional production, which must be transferred as output from at least one supplier to a recipient and induces a flow of money that gives rise to additional purchasing power. The increase in value added thus has a material origin side and a financial use side of the income increase induced by it. Both effects would have to be ecologically neutralized in order for the economy to grow in an ecologically harmless way. In other words, even if it were ever possible to technically dematerialise the creation of a monetary and thus GDP-relevant transfer of benefits—which, with the exception of singular and hardly scalable laboratory experiments, is not yet foreseeable—the decoupling problem would remain unsolved as long as the additional income could be used to finance any goods that are not completely dematerialised. Both decoupling problems will be briefly examined using the example of the energy transition.

4.7.1 Output Side of GDP: Material Rebound Effects

What would goods have to be like that are transferred from at least one supplier to a consumer as monetary services, but whose production, physical transfer, use and disposal are free of all land, material and energy consumption? Green Growth solutions devised so far obviously do not meet this requirement, regardless of whether they are passive houses, electric vehicles, wind turbines, photovoltaic systems, combined heat and power plants, smart grids, solar thermal heating systems, car sharing, energy-saving light bulbs, digital services, and so on. None of this is possible without physical input, in particular new production capacities, distribution systems, mobility and the necessary infrastructures, which must therefore lead to further material addition as long as economic growth is to be fed from it.

But could the "green" solutions not simply replace the less sustainable production instead of being added together, so that on balance there is an ecological relief? No, because firstly, it would not be sufficient to replace only output flows, as long as the structural change inevitably required for this would be accompanied by an addition of material stock sizes and land consumption (as in the case of passive houses or renewable energy installations). Consequently, existing capacities and infrastructures would have to be eliminated. But how could the material of entire industries, building complexes or several millions of fossil-fuelled cars (to replace them with e-mobiles) and heating systems (to replace them with electric or solar thermal systems) disappear without entropy production?

Secondly, GDP would not be able to grow systematically if every gain in green value creation were offset by a loss due to the dismantling of old value creation structures. Thus, the value-added contributions of renewable energies, which are currently being hailed as the flagship of a prosperous "green economy", turn out to be a flash in the pan on closer inspection. Once the temporary phase of capacity expansion is over, the contribution to value creation is reduced to an energy flow that is likely to have only comparatively modest effects on GDP and the labour market and could only be increased if the construction of new plants were to continue indefinitely. But then there would inevitably be additional environmental damage: the material stock sizes would expand and the destruction of the landscape, which is already barely acceptable, would increase accordingly.

This highlights an insoluble dilemma: to the extent that even "green technologies" can never—certainly not when all system requirements are considered holistically—be immaterial, i.e. available at zero ecological cost, their theoretically maximum relief effect consists in any case only in transforming environmental damage into other manifestations or shifting it into other ecological media instead of avoiding it. This is done in four ways.

1. The physical shift can be demonstrated by the example of the energy-saving light bulb, which is more energy-efficient than the standard light bulb but proves to be more problematic in terms of production and disposal.
2. Spatial shift effects consist of moving environmentally intensive process stages of manufacturing to distant countries (often China or India), so that the ecological damage is no longer recorded in Europe's environmental balance sheets.
3. Some environmental technology innovations, such as composite thermal insulation systems or photovoltaic systems, turn into a disposal problem after about 20 years, so that there is a time lag here.

4. Other measures, such as wind turbines, produce comparatively fewer emissions (they cannot be completely emission-free due to the production of the turbines), but they consume or impair landscapes and land. This is a systemic shift, i.e. environmental damage is transferred from one physical aggregate state to another, but not avoided. Moreover, such a shift will reach quantitative system limits at some point, for example when all suitable areas are occupied.

Presumably environmentally beneficial innovations can even cause several of the above-mentioned displacement effects. Therefore, the attempts to empirically prove ecological relief successes of the energy transition are pointless, especially since the displacement effects cannot be converted into CO^2 equivalents. Even if CO^2 savings worth mentioning were to be achieved at some point, how many hectares of impaired or destroyed landscapes would a society affected by dramatic land scarcity be prepared to pay as a price for this? And even if this price were accepted, it would hardly be possible to speak of progress in sustainability, insofar as only one particular category of damage would be exchanged for another. This finding reveals an ambivalence of the technical progress that is supposed to decouple economic growth from environmental damage: More in the here and now is bought with less elsewhere and later.

4.7.2 Demand Side of GDP: Financial Rebound Effects

Even if it were possible to dematerialise increases in production, the increases in income that inevitably correspond to growth would also have to be ecologically neutralised in order to achieve relief effects. But it is simply inconceivable to keep the shopping basket of those consumers who draw the additional income generated in the supposedly green sectors free of goods whose globalised production involves fossil energy and other raw materials. Would people employed in the green industries not build homes, travel by air, drive cars, or engage in the usual consumer activities—and with an upward trend as disposable income grows?

A second financial rebound effect threatens if green investments increase total output because the old production capacities are not dismantled at the same time and to the same extent (the total living space increases due to passive houses, the total amount of electricity increases due to photovoltaic systems), which tends to cause price reductions and consequently increases demand. Regarding the electricity sector in particular, we refer to Table 2.5. Here, the growth of gross electricity generation by gas-fired power plants, biomass use, wind and photovoltaic plants can be seen. The fact that the price of electricity in Germany did not fall for households is due to the special features of the energy transition, which are discussed in Sect. 5.1.1. However, this does not change the fundamental capacity problem of the green growth paradigm. A third financial rebound effect, already described by Jevons [149], occurs when efficiency increases reduce the operating costs of certain objects (houses, cars, lighting, etc.).

Theoretically, these financial rebound effects could be avoided if all income increases were siphoned off, but then what is the point of growth at all: what could be more absurd than generating growth in the first place only to neutralize the income increases intended by it? So the claim that green technologies can bring about economic growth accompanied by an absolute reduction in environmental pollution is not only false, it is the exact opposite: From the perspective of financial rebound effects, only in case of a non-growing GDP do green technologies have any chance at all of relieving the ecosphere. And this is not even a sufficient condition, because the material effects, in particular the innumerable displacement possibilities on the generation side, must also be taken into account.

4.8 Post-growth Economy

Insofar as plunder-free economic growth would be tantamount to squaring the circle, self-limitation alone remains as the way out. This corresponds less to an ethical imperative than to mathematical logic. The industrial value creation and fossil mobility would have to be reduced in consumer societies in such a way that per capita resource consumption falls to an ecologically transferable level. The amount of the reduction required to achieve this is easy illustrated by the widely accepted two-degree climate target: if the total amount of CO^2 emissions compatible with this target were to be distributed evenly globally among some 8 billion people, this would result in a maximum individual budget of around one ton per year.[1] According to the Federal Environment Agency [151], this measure is actually around 12 tonnes in Germany.

4.8.1 Five Milestones of a Transformation

What would the counter-design to ecological modernisation look like? Instead of optimizing an irredeemable industrial model, its scope and volume would have to be dosed. Making the dismantling of industrial supply socially acceptable and economically resilient lies at the heart of the post-growth economy. This requires various development steps that address both the demand and the supply side.

1. Sufficiency: Tapping reduction potentials on the demand side is not synonymous with renunciation. The sufficiency principle confronts excesses of self-fulfilment

[1] In 2009, the German Advisory Council on Global Change (WGBU), working on behalf of the German government, estimated the maximum global budget of CO emissions available up to 2050 that would be compatible with compliance with the two-degree climate target, and distributed it among the world's 7 billion inhabitants at that time. This resulted in a rough guideline value of 2.7 t per capita and year until 2050, after which this value would have to be reduced again [150]. Since not only the world population has increased by at 1 billion in the meantime but also the concentration of carbon emissions, a correspondingly corrected guideline value of 1 t seems plausible.

with a simple counter-question: From which energy slaves and comfort crutches can over-exuberant lifestyles and the Society as a whole be liberated for its own benefit? What junk of prosperity, which has long since clogged up life and, on top of that, takes up time, money, space and ecological resources, could be gradually phased out? A "time-economic theory of sufficiency" [152] provides reasons for this beyond moral appeals. In a world of information and option overload that no one can process any more, deceleration becomes psychological self-protection. The increasingly "exhausted self", as depicted in Ehrenberg's classic of the same name [153], embodies the shadow side of a merciless pursuit of happiness that increasingly turns into overload. Liberation from excess would mean limiting oneself to a selection of consumption activities and objects that can still be managed, given limited attentional resources. Self-limitation and, above all, sedentariness—global mobility has long since displaced consumer activities as the most climate-damaging form of self-realization—form a prerequisite for an art of living that is both responsible and enjoyable.

2. Subsistence: Consumers could acquire competencies that would enable them to satisfy some needs on their own, beyond recourse to commercial markets. If industrial production were to be sufficiently reduced, the resulting reduction in the amount of wage labour time still required could be redistributed in such a way that full employment would be possible on the basis of 20 h of working time per week. This would free up time resources for self-sufficiency. Community gardens, exchange rings, networks of neighbourly help, give-away markets, facilities for the communal use of equipment and tools would not only lead to a gradual de-globalisation, but also to a reduced need for technology, capital, transport routes and, moreover, to more autonomy. If products are used for longer, maintained, repaired and cared for independently, and if possible acquired second-hand, the dependence on industrial supply is reduced. The shared use of consumer goods has a similar effect. A longer useful life and increased number of users of the same object reduces the need for material production and income to finance all necessary expenditures.
3. Regional economy: Consumption needs that can be reduced neither by sufficiency nor by subsistence can be satisfied on regional markets with greatly shortened value chains. Complementary regional currencies to be introduced in parallel with the euro could tie purchasing power to the region and thus decouple it from globalised transactions. In this way, the efficiency advantages of a money-based division of labour would still be exploited, but within a small-scale, more ecologically compatible and more resilient framework. In particular, in food production (e.g. Community Supported Agriculture), community use and extension of useful life, regional economic enterprises could operate where the potentials of subsistence end.
4. Restructuring of the remaining industry: The residual demand for supra-regional industrial value creation would focus on the optimisation of already existing objects, namely through refurbishment, renovation, conversion, redevelopment and intensification of use in order to provide utility services as production-free as possible. Markets for used and repaired goods as well as commercial sharing and

rental systems also contribute to this. The new production of material goods would be limited to maintaining a constant stock of material goods, i.e. only replacing what can no longer be maintained through efficient life extension. In addition, the manufacture of products and technical devices would be oriented towards a repairable and both physically and aesthetically durable design.

5. Institutional measures: Conducive framework conditions include land, monetary and financial market reforms, with the financial transaction tax (Tobin tax) and a wealth tax being highlighted. Each person would be entitled to an annual CO_2 emission quota of 1 t, which could be transferable over time. Changed forms of enterprise such as cooperatives, non-profit organisations or concepts of solidarity-based economic activity could dampen profit expectations. Subsidies—especially in the areas of agriculture, transport, industry, construction and energy—would have to be eliminated in order to reduce both the ecological damage they cause and the public debt. Measures that facilitate reductions in working hours are indispensable. There is also an urgent need for a moratorium on soil sealing and deconstruction programmms for industrial sites, motorways, car parks and airports, in order to unseal and renaturalise them. Otherwise, renewable energy plants could be built on disused motorways and airports to reduce the catastrophic landscape consumption of these technologies. Furthermore, precautions against planned obsolescence are essential. A drastic reform of the education system would have to aim at teaching manual skills in order to be able to reduce the need for new production through self-production and above all maintenance and repair measures to become more independent of money.

4.8.2 Reducing Growth Imperativs

The categories of measures roughly marked in Sect. 4.8.1 are not only derived from the need to reduce the output of material goods and fossil mobility, but are also intended to stabilize a certain level of production, i.e. to eliminate growth constraints.

The industrial system is characterised by functionally highly differentiated, i.e. spatially delimited and complex value chains. If service production, which was previously tied to a production location, is broken down into as many isolated production stages as possible, this allows them to be relocated flexibly. Each isolated sub-process can be moved to where costs are minimized through specialization and economies of scale. Consequently, wealth creation requires a growing number of intermediate levels of specialization. Through this differentiation, it is possible to standardize the respective design of the individual tasks and modules. In this way, processes can be automated, i.e. human labour can be replaced or reinforced by technology that converts energy and matter, which is commonly understood as an increase in labour productivity.

Each participating company must pre-finance the required inputs before starting production, i.e. invest, which requires debt and/or equity capital. But this cannot be obtained at zero cost. Therefore, companies must generate a surplus in order to be able to finance the interest on debt and/or return on equity. The more capital-intensive production is as a result of increasing specialisation, mechanisation and the number, size and geographical distance of the work stations involved, the higher, ceteris paribus, is the surplus necessary to satisfy the demands of the capital owners. This results in a minimum amount for the total production growth necessary to stabilise the value creation process.

Binswanger [145] has analysed this structural growth constraint in connection with the income and capacity effect of an investment. The income effect starts before the capacity effect, because capital is invested first and a sale of the production quantity is only possible afterwards. Investment made today immediately increases household income. But the quantity of output resulting from the investment can be sold only later, in the subsequent period. Therefore, households buy today yesterday's output. In this way, the increase in demand precedes the increase in supply. If, on the one hand, expenditure precedes income, but, on the other hand, both are expressed in the form of money payments, the difference between which corresponds to profit—how can this ever be positive within a period? This is only possible if the "payment gap" on the demand side is compensated by additional net investments, which create the corresponding income. But this induces at the same time a capacity effect, which increases the quantity of production, which in turn must be sold subsequently in order to cover the costs and the claims of the shareholders.

Insofar as this logic applies to every specialized enterprise that requires capital in order to produce, starting points for a mitigation of structural growth constraints can be derived: Fewer specialisation stages between production and consumption reduce the economic cost advantages of the division of labour, but can also reduce the growth imperative, insofar as the capital intensity of production and consequently the sum of the minimum surpluses to be achieved in order to satisfy the capital claims is reduced. Short value chains, for example in the sense of a local or regional economy, also create proximity and thus trust, which per se can make it possible to raise capital at lower interest rates and with lower returns. This is because both equity and debt investors bear an investment risk that increases due to technical and spatial complexity of supply chains. With this risk grows the financial compensation demanded by the providers of capital.

Another side effect of capital-intensive value creation processes is that the technical progress exploited through them constantly increases labour productivity. For this reason, any level of employment once achieved can only be maintained after a surge in innovation if the production volume grows sufficiently. Overall, starting points for reducing structural growth constraints extend to

- the combination of different value-added systems in order to directly influence capital or labour intensity,
- Self-sufficiency practices to reduce the need for industrial production,
- technologies, which per se correspond to a higher work intensity,
- Reduction and redistribution of working time,
- Sufficiency of consumption patterns and
- short value chains to reduce expected returns on capital or interest.

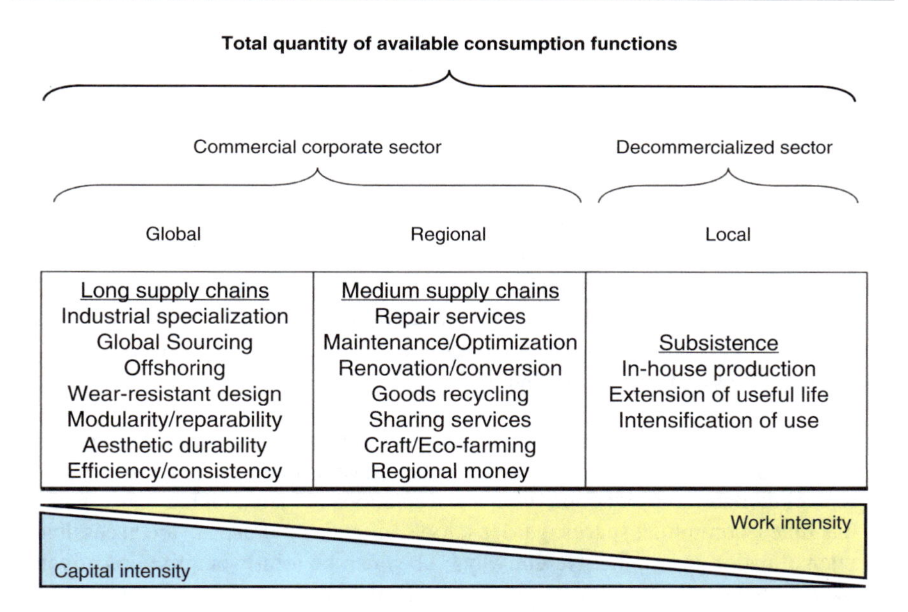

Fig. 4.2 Supply systems of the post-growth economy

The first three of these factors in particular will be addressed below.

4.8.3 Industrial Supply and Subsistence

First of all, three ideal-typical supply systems can be distinguished: (1) global industrial division of labour, (2) regional economy and (3) modern subsistence. The transformation to a post-growth economy would correspond to a structural change which, in addition to exhausting all reduction potentials (sufficiency), would gradually and selectively shift the remaining production from the first to the second and third aggregate. The three systems are characterised by different capital intensities. Moreover, they complement each other and can be synergistically linked to form a modified value creation structure—especially the first and third aggregates (Fig. 4.2).

End users, who only have the role of (passive) consumers within conventional value creation processes, can contribute to the substitution of industrial production as "prosumers" [154]. Prosumers are individuals who, in addition to regular employment, make their own productive contributions to the satisfaction of their needs. In this context, new balances between self-sufficiency and industrial supply are possible, which can take a wide variety of forms. Between the extremes of pure subsistence and global interdependence there is a constant continuum of various degrees of external supply. The supply through industrial production can be replaced by one's own production, either selectively or gradually.

Self-sufficiency practices unfold their effects in the immediate social environment, i.e. at communal or regional level. They are based on a (re-)activation of the competence to satisfy needs manually and by means of one's own activities beyond commercial markets, above all by means of manual skills, which are less energy-intensive but more time-intensive. The time required for this would result from a dismantling of the industrial system, combined with a redistribution of the regular employment still required. By reducing average weekly working hours, self-supply and external supply could be combined in such a way that the supply of goods would be based on a lower monetary income, but supplemented by market-free production (subsistence). In addition to voluntary, community-oriented, educational and artistic activities, modern subsistence comprises three output categories through which industrial production can be gradually substituted:

1. Intensification of use: Borrowing a commodity from a neighbour, baking him a loaf of bread in return, or installing the latest Linux update, contributes to replacing material production with social relations. Objects such as cars, washing machines, communal spaces, gardens, tools, digital cameras, etc. are accessible to intensification of use in different ways. They can be jointly acquired or privately owned by a person who makes the object available in return for other subsistence services. In some cases, so-called *commons* (common property), which have been extensively researched by Ostrom [155], are suitable as an adequate institution. This concerns resources, goods or infrastructures which, similar to a commons, are used jointly by a defined group of people based on certain rules.
2. Extending the useful life: Special importance would be attached to the care, maintenance and repair of consumer goods of all kinds. Those who increase the useful life of consumer objects through craftsmanship or manual improvisation skills—sometimes even careful treatment is enough to avoid early wear and tear—replace production with their own productive services without necessarily foregoing previous consumption functions. If it were possible in a sufficient number of consumer goods categories to double the useful life of objects on average through maintenance measures and repair, the production of new objects could be halved accordingly.
3. Home production: In the food sector, home gardens, roof gardens, community gardens and other forms of urban agriculture are proving to be a dynamic trend that can contribute to the de-industrialisation of this sector. In addition, artistic and handicraft achievements are possible, ranging from the creative recycling of discarded objects to one-off wooden or metal objects [156].

Inasmuch as a post-growth strategy entails not only a decline in industrial production but also, inevitably, a reduction in income, subsistence activities are needed to ensure resilient, i.e. crisis-resistant, supply systems. This applies in particular to basic needs, especially food. This raises the question of the efficiency of local and regional food supply. There is still a need for research on the question of how high the potential share of own production in the food supply could be. A study [157] prepared at HafenCity University in 2016 used the example of the city of Hamburg

Table 4.1 Regional supply using the example of Hamburg

	Style 1	Style 2	Style 3
Features of the eating style	Status quo, conv.	Kattendorfer Hof, organic	Status quo, bio
Meat/head	87 kg	36 kg	87 kg
Area/head	2388 m^2	2346 m^2	3102 m^2
Region 1 (Hamburg)	3%	3%	3%
Region 2 (50 km)	48%	49%	37%
Region 3 (100 km)	97%	99%	75%
	Style 4	Style 5	Style 6
Features of the eating style	DGE organic	−30% meat organic	DGE organic, veget.
Meat/head	24 kg	61 kg	0 kg
Area/head	2054 m^2	2802 m^2	1936 m^2
Region 1 (Hamburg)	4%	3%	4%
Region 2 (50 km)	56%	41%	60%
Region 3 (100 km)	100%	92%	100%

Meat/capita = annual average consumption, DGE = German Nutrition Society.

to explore the potential of supplying a metropolis with food completely locally/regionally. Although the result (see Table 4.1) does not directly address the capacity of community gardens and other forms of urban agriculture, it emerges firstly that it is possible to supply 100% of Hamburg's food within a radius of 100 km even on the basis of organic farming, and secondly that the resource requirements (especially land) are critically dependent on the dietary style. In other words, the more sufficiency-oriented the dietary culture, the more favourable the possibilities for local supply.

The forms of subsistence described above—especially the extension of useful life and communal use—can have the effect that a halving of industrial production and consequently of labor employment does not per se halve material prosperity: When consumption objects are used longer and are shared among users, a fraction of current industrial production is enough to extract the same quantum of consumption functions inherent in those goods. Subsistence, then, consists in upgrading or "refining" a markedly reduced industrial output by adding one's own inputs. These subsistence inputs can be classified into the following three categories:

1. Craftsmanship and improvisational skills to exploit the potential of self-production and service life extension.
2. Own time which must be spent in order to be able to perform manual, substantial, manual or artistic activities.
3. Social relations, without which subsistent community uses are unthinkable.

Urban subsistence is the result of a combination of several input and output categories. Let us assume that prosumer A has a defective notebook repaired by prosumer B, who has the corresponding skills, and in return gives him organic carrots from the community garden in which he participates. Then this transaction is based firstly on social relations that person A enters into with both B and the garden community, secondly on manual skills (A: growing vegetables; B: renewing a defective hard drive and installing a new operating system) and thirdly on one's own time, without which both manual activities cannot be performed. The outputs extend to own production (vegetables), service life extension (repair of the notebook) and community use (garden community).

Of course, subsistence actions are also obvious, which do not require exhaustion of the full range of conceivable subsistence inputs and outputs. Those who cultivate their own garden, increase the useful life of their textiles by repairing them themselves, or look after their children themselves instead of consuming all-day care are not using social relations, but they are using time and manual skills. In this example, the outputs extend to the extension of useful life and self-production.

Insofar as subsistence combinations in the above sense replace industrial output, they simultaneously reduce the need for monetary income. A necessary condition for achieving lower levels of external supply thus consists in a synchronization of industrial reduction and compensating subsistence accumulation. In this way, a reduction in monetary income and industrial production could be socially compensated, although not at the previous average material level. Therefore, this transition is not conceivable without the logic of sufficiency, which will not be discussed in detail here.

4.8.4 Supply Configuration

In contrast to the traditional concept of subsistence, the self-sufficiency practices described above are closely intertwined with industrial production. In particular, de-commercialised extension of useful life and intensification of use can be understood as non-industrial, consequently capital-free extension of supply chains. This is done by adding non-market and independently provided inputs. In contrast to traditional economics, the supply of goods (hereafter referred to as v) can be represented as the availability of service and use units, which depends not only on industrial production, but also on the activity level of subsistence i. This is fed firstly by the intensity and duration with which a given stock of industrial goods is used, and secondly by supplementary services of self-production (for example through community gardens).

This subsistence factor or output i can be understood as a function of the above-mentioned inputs, namely time t spent by the individual, manual competence h and social relations for the purpose of exchange of services or division of labour s, in other words

$$i = w_{h,s}(t). \tag{4.1}$$

This quantity describes the yield or intensity with which industrial production Y is used. The latter is written as in Eq. (3.1) as

$$Y = F(K, L, E; t), \quad (4.2)$$

where $K = K(t)$ denotes the capital employed, $L = L(t)$ the labour input and $E = E(t)$ the energy input. Material wealth thus becomes more independent of industry, especially since it is based on two sources:

$$v = Y + i = F(K, L, E; t) + w_{h,s}(t). \quad (4.3)$$

Consequently, the production process changes: industrial production is coupled with the subsistence production that complements it. Production, use and subsistence—the latter understood as activities that preserve and enhance the stock of objects—complement each other to form a multi-component value creation process. Prosumers contribute independently to the preservation of their stock of goods, thereby reducing the necessary replacements and thus overall the need for industrial production. The latter could also be seen as a preliminary stage of value creation for subsistence activities.

The integration of creative subsistence services gives rise to a cascade-like value creation structure. It extends to careful use, care, maintenance, servicing, modular renewal and independent repair work. This is followed by the reuse of dismantled components and, if necessary, adaptation to other uses. The latter includes "upcycling" practices, the assembling of individual parts of several objects that are no longer functional into one usable object. This is followed by the storage, sale or transfer of dismantled parts to collection points and repair workshops. In addition, there is the possibility of passing on goods that are still fully functional to so-called giveaway markets or "free shops".

This usage cascade has further interfaces to commercialized usage and production systems. Both functioning products and dismantled individual parts or modules can be sold via second-hand retailers, flea markets or Internet-based intermediaries (eBay, Amazon Marketplace, etc.). Maintenance and repair work, which would overburden prosumers, can be carried out by professional craftsmen. These would be part of the regional economy. Their role would also be to substitute productive services of the industrial sector on the basis of production methods that tend to be less capital- and energy-intensive and shorter ranges of the value chains.

While the industrial sector is characterized by a relatively high energy and capital intensity, the value added of the subsistence sector is almost exclusively fed by non-market goods. With regard to the entire process chain, the average energy and capital intensity per unit of utility is thus reduced. Instead, the labour intensity increases, which means that the productivity of the factor labour decreases—but only in relation to the entire process, consisting of industrial production and the (labour-intensive) subsistence production linked to it. The higher labour intensity need not therefore concern industrial production, which can continue to be characterised—but with a reduced output quantity—by specialised and relatively capital-intensive manufacturing processes. Rather, it follows from an artisanal extension and intensification of product use, so that the capital employed in the

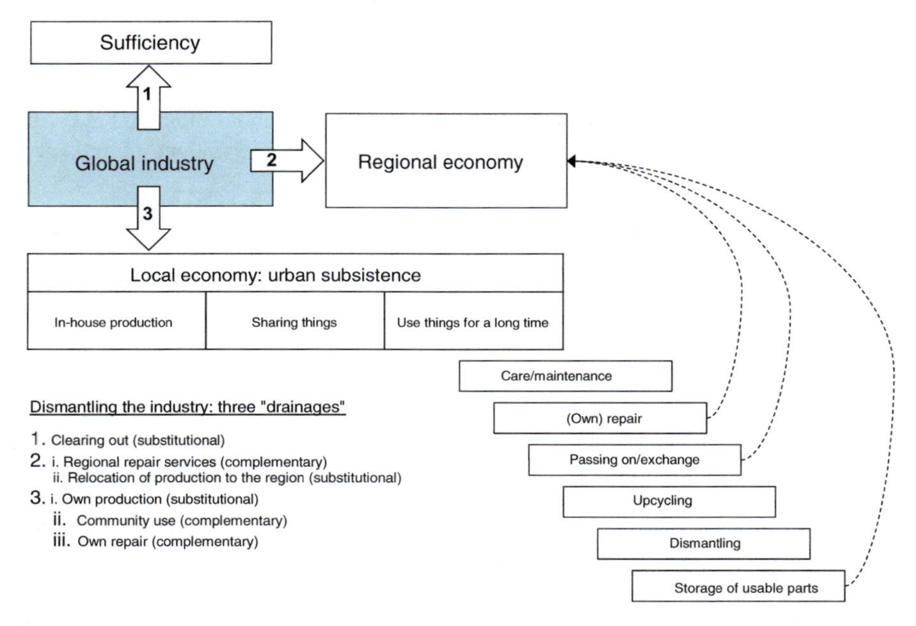

Fig. 4.3 Starting points for reducing industrial output

industrial segment is distributed over a more productive quantum of potential uses. Therefore, customers can contribute to reducing the absolute level of the required capital stock in two ways, namely by being sufficiency-oriented enough to orient themselves towards available options for use rather than towards the ownership of goods, and by contributing subsistence capabilities (Fig. 4.3).

In addition, there is a substitutional relationship between the two sectors. This is based on the fact that independent production, for example through community gardens, handicraft or artistic production, leads to the direct substitution of industrial products. The relationship between subsistence and regional economy can also be both complementary, as already outlined above, and substitutional. The same applies to the transformational relationship between industrial and regional value creation. A complementary relationship arises where regional, craft-oriented businesses add value to reduced industrial output through repair and maintenance services. In addition, industrial goods can be gradually substituted by regional production (at least to a greater extent than at present) (food, textiles, certain consumer goods, etc.).

4.8.5 Adapted Tools to Reduce Capital Intensity

Both substitutional and complementary transitions from the industrial sector to subsistence and regional economies can be further reinforced by changing technological conditions. Kohr [158] distinguishes between primitive, intermediate and advanced technologies, each corresponding to a corresponding size of the relevant social system or society. The intermediate technologies he favors are not only less complex, but avoid a limitless and unconditional maximization of labor productivity. The same applies to "convivial" technologies described by Illich [159]. According to this, there would not be a complete substitution of physical labour by external energy supply and capital input. Rather, the aim is to achieve a balance between manual tasks and their reinforcement by means of moderate energy input. Like Kohr, Schumacher [141] also emphasizes the decentralized aspect of medium technologies.

The low capital intensity of such "amplifiers of human power", as Illich puts it, can have the effect that their availability does not depend on high investment sums. Thus, intermediate or convivial technologies have a basic democratic and socially levelling character. Their availability presupposes neither wealth nor power. Schumacher associates this with the shift from mass production to the "production of the masses". The idea of emancipation implied by this has recently been taken up by Friebe and Ramge [156]. While they object to the "return to a pre-industrial craftsman's idyll", a brief review of this stage of development certainly proves instructive.

Mumford [160] characterizes technologies used before industrialization as follows:

> Although they worked slowly, trade and agriculture before mechanization, precisely because they were based primarily on manual labor, possessed a freedom and flexibility unlike any system that depends on a set of costly specialized machines. Tools have always been personal property, chosen to suit the needs of the particular worker and often redesigned, if not specially made. Unlike complex machines, they are cheap, replaceable, and easily transported, but worthless without human power.

Another feature of adapted technologies is their shorter spatial reach, i.e. shorter distances between consumption and production. This results not only in a high degree of compatibility with approaches to subsistence and regional economy, but also in the possibility of their independent design and repair. Such technologies are flexible, controllable and can be used autonomously. On this basis, more resilient forms of supply and existence are possible. They not only protect against exclusion and manipulation, but also guarantee stability. Insofar as a flexible "polytechnic" [160, p. 487 ff.] takes the place of unifying and central structures, a variety of tools options. This firstly contributes to crisis resistance (resilience) and secondly keeps open a richer variation of development paths and possible reactions to crises.

The various varieties of adapted technologies empower prosumption, without which a post-growth economy hardly seems possible. Moreover, in addition to the subsistence effect outlined in Sect. 4.8.4, they contribute to a reduction in capital intensity, which not only implies less pressure to exploit capital, but also helps to stabilise a certain level of employment without or at least with lower growth rates. Another criterion concerning the dependence on (expert) knowledge is emphasized by Illich [159, p. 91]: "How much someone can learn on his own depends very largely on the nature of his tools: The less convivial they are, the more training they require." Adapted technologies would thus liberate not only from a monopolization of indispensable knowledge, but from the constraints and exclusionary tendencies of the knowledge society. Their democratic character, their financially unconditional availability, and their individualizability help to cushion the necessary deconstruction of industry in social terms.

4.8.6 Material Zero-Sum Games

The rebound effects mentioned in Sect. 4.7 result not least from an unquestioned orientation towards innovation in the sustainability discourse. In the Green Growth context, it is mostly ignored that innovation as a specific mode of change is neither neutral nor without alternative. Innovation processes have an expansive character per se, because they add additional options to the pool of existing possibilities—even if, viewed in isolation, they are comparatively more sustainable solutions. Even optimised goods and technology variants that may be ecologically efficient, recyclable or renewable can never manage entirely without energy, land and minerals. If they are added to the current stock of goods, (material) growth inevitably results—precisely because too much new is entering the world and too little old is disappearing.

Creating more sustainable alternatives as a result of innvoation conjures up considerable risk problems in addition to rebound effects [139, p. 230 ff.]. Moreover, for many ruinous practices there are simply no functionally adequate substitutes. A transformation strategy that is conditional on creating an equivalent substitute for every unsustainable mobility, consumption or production pattern, rather than eradicating the causes of damage without substitution if necessary, amounts to the well-known failure of ecological modernization. From a growth-critical perspective, sustainability is not so much an endeavour of additional effect as of careful omission. Consequently, the chronically expansive innovation principle needs to be supplemented by contractive forms of design, referred to as "exnovation" [139, p. 259 ff., 303 ff.], in order to treat the addition and departure of elements of the possibility space as equally valid options.

While innovation processes are oriented towards the question "How does the new come into the world?", the exnovation principle is complementarily oriented towards the challenge "How does the old, formerly innovative, but which has meanwhile become a problem, come out of the world again without harm?". Examples of exinnovation processes include ending the manufacture of asbestos-containing eternit products or CFC-based refrigeration equipment.

Beyond expansion and contraction, renovation as a third design principle internalizes change within a given option space. Instead of increasing or decreasing the stock of goods, existing objects are upgraded through refurbishment, maintenance, repair, functional expansion, and provided with additional potential uses through consumer goods recycling [139, p. 455], especially efficient forms of second-hand trade. In addition to subsistence, professional renovation strategies

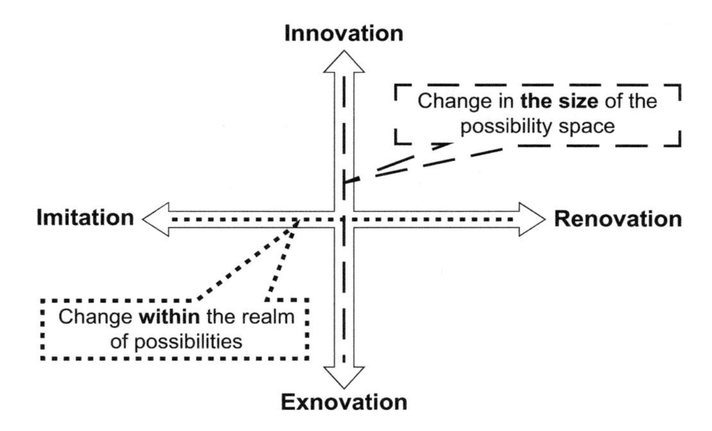

Fig. 4.4 Different modes of change

contribute to extracting increased benefits from existing goods and infrastructures by adapting them functionally and aesthetically to current needs (Fig. 4.4).

The interplay of these three principles (imitation will not be discussed here) can be used to form entrepreneurial strategies by which the industrial system can not only be reduced to an ecologically transferable level, but can also be redesigned on the production side in such a way that material flow and stock levels remain at a constant level. The focus would be on stock maintenance in those areas with which prosumers (cf. Sect. 4.8.3) would be overwhelmed. New production could be minimised if it were restricted to the replacement of those objects that cannot be maintained in a efficient or sensible way. Daly [161] calls such a state a *steady state economy*.

Linked to this is the concept of so-called material zero-sum games [139, p. 455]. These embody the physical-material dimension of a growth-neutral supply of goods. In this context, additional material flows or the addition of material stock sizes are considered only as a last resort, to be avoided or minimized as far as possible. Where necessary, additive and reductive measures should balance each other in such a way that any production serves only to maintain, but precisely not to expand, a constant stock of material artefacts. This encompasses two perspectives:

- In order to keep the extent of material and energy flows largely constant, new uses are wrested from the pool of already produced objects and occupied areas. These include not only utilization systems for the production-free satisfaction of needs, but also services that serve to upgrade, repurpose, recombine, convert or optimize the use of existing consumption and production hardware. Growth-neutral changes focus on a careful transformation of the already occupied ecological space instead of bringing new material artefacts into the world. Accordingly, for example, the energetic renovation of an old house would be preferable to the construction of a new passive house, no matter how refined.

- If material objects are added and additional ecological capacities are used, this must be accompanied by a compensatory subtraction elsewhere. For example, any further sealing of land would have to be accompanied by compensatory unsealing. The same applies to the replacement of an object that can no longer be used. Mere replacement at the end of a service life that can no longer be (sensibly) extended does not increase the stock, but merely maintains it. For the rest, the replacement may consist in a more efficient or consistent solution.

When the approach of the material zero-sum game is combined with the three modes of change and different levels of design (product, process, service, system of use, organization, institution), a four-stage search corridor emerges:

1. Direct link between innovation and exnovation. Example: Product innovations grant a high degree of growth neutrality if they do not generate any new consumption needs, but fulfil the existing ones more efficiently or more consistently, so that neither a motivation for premature discontinuation nor for parallel acquisition is aroused. This merely results in the replacement of products whose useful life can no longer be extended.
2. Direct link between innovation and renovation. Example: Insulating materials made from renewable raw materials (product innovation) can be used for the thermal insulation of old buildings (product renovation).
3. Indirect link between innovation and renovation. Example: Certain service innovations such as maintenance, refurbishment or repair can contribute to increasing the useful life or intensity of the existing product stock (product renovation). Institutional innovations such as the establishment of effective intermediaries for the trade in second-hand goods can also enable the renovation of consumer goods.
4. Indirect link between innovation and exnovation. Example: Approaches to car sharing as a system and service innovation can have the effect that previous owners of a car do not purchase a new vehicle after it has been discarded (product exnovation), but instead demand mobility services.

These coupled strategies are illustrated in Fig. 4.5. A direct link between two modes of change means that they are applied at the same design level, for example when a *product innovation* and a *product exnovation* materially cancel each other out. An indirect link between two modes of change, on the other hand, means that they are applied at different levels of design. An indirect link between innovation and exnovation could mean, for example, that a *service innovation* replaces material objects, i.e. is linked to a *product exnovation*. The above innovation strategies can be used to structure entrepreneurial search processes.

Avoiding economic growth suggests a prioritising ranking of the four search fields. Accordingly, for a specific need, solutions should first be sought that are not based on material objects but on services (option 4). If product-replacing services in the relevant field of need are not compatible with market or cultural conditions, solutions of this kind could overstretch the company's design potential. In this case,

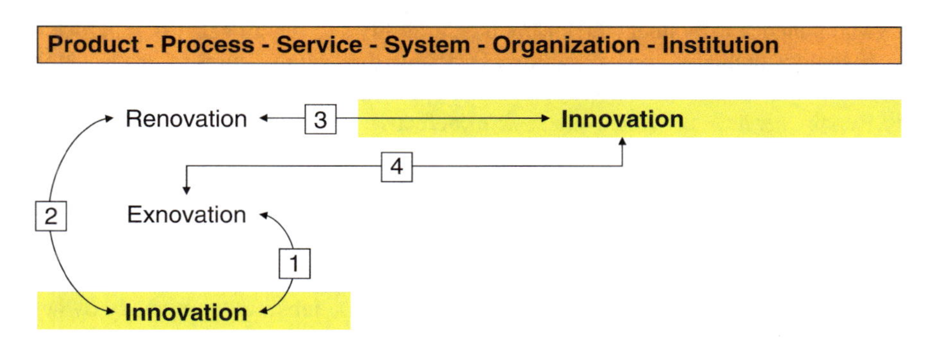

Fig. 4.5 Coupled innovation strategies

the next field of action would be to look for possibilities that include product ownership (option 3). This would involve services that maintain the material stock, i.e. that extend or intensify the stream of benefits to be derived from it. If this search field does not reveal any suitable alternatives either, product innovations come into consideration, initially as a minor addition to an existing object for the purpose of upgrading or improving efficiency (option 2). Only as a last strategy is a conventional product innovation applied, but coupled with an exnovation that occurs after all possibilities of a service life extension have been exhausted (option 1).

4.8.7 Growth-Neutral Business Areas

From the foregoing, it can be concluded how companies can contribute to a post-growth economy:

- Shortening value chains and strengthening creative subsistence.
- Measures that facilitate a reduction and redistribution of working time to feed the subsistence input "own time".
- Local and regional sourcing to unbundle supply chains.
- Support for and participation in regional currency systems.
- Direct and regional marketing.
- Developing modular, repairable, recyclable and physically and aesthetically durable product designs to facilitate urban subsistence services, moving away from "planned obsolescence".
- Prosumer management: Beyond pure production companies could offer courses or training to enable users to maintain, service and repair products themselves.

Companies that follow the logic of zero-sum material games would be recognizable, among other things, as

- Maintenance personnel who ensure the durability and functionality of a hardware inventory that is as non-expanding as possible through measures of preservation, maintenance, preventive wear reduction and consulting,
- Repair service providers that save defective goods from being discarded prematurely,
- Renovators who extract further value from existing assets by adapting them functionally and aesthetically to current needs,
- Redesigners who recombine, convert or repurpose existing infrastructures and hardware in such a way that new uses emerge from them,
- Providers of services that, in appropriate situations, substitute previously owner-occupied forms of consumption with services,
- Intermediaries who, by reducing the transaction costs of the trade in second-hand goods, ensure that consumer and capital goods are kept in the cycle of efficient use for as long as possible,
- Designers who orient a smaller quantity of newly produced material objects towards durability and multifunctionality as well as
- Skills facilitators who help turn consumers into prosumers through training, practical courses and the provision of workshops.

Taking these possibilities into account, the dogmatic history of entrepreneurship could be reconstructed from a different angle and continued in a post-growth-compatible manner: When people began to live urbanized lives, it became necessary to ensure a supply that could no longer be provided by subsistence alone, namely through organizational forms that exploited the advantages of the division of labor. This heralded the first phase of entrepreneurship. The generation of material surpluses, accompanied by and based on specialized labor, resource pooling, market exchange, money, consumer cultures, technical and institutional innovations, culminated in a sequence of industrial revolutions.

A second stage in the development of entrepreneurship could be associated with an accent on services, experiences and symbols, admittedly without pushing back the material sphere that continues to exist in parallel, but on the contrary even spurring it on. The next phase, which is far from being fully developed, could be an economy of maintaining and upgrading existing assets. Accordingly, companies would hardly produce anything new, but rather maintain, repair and refine the pool of goods that have long existed, in order to creatively wring new potential uses out of it.

Finally, a fourth developmental step could be a return to subsistence, albeit with a modern face. Having always promoted the spread of consumer cultures, companies could move not only to substitute services and maintenance for the physical production of goods, but also to support self-sufficiency practices. Companies that empower consumers to become prosumers help them not only to practice an ecologically transferable style of care, but also to achieve greater economic autonomy.

Suppose that the purchase of a computer included the use of a prosumer training course, so that buyers were taught the necessary basic knowledge and skills to independently renew modules or repair possible predetermined breaking points that remain despite a durable design. Then the average useful life could be doubled or tripled independently and manually, and the need for new purchases halved or tripled. Such new compositions of entrepreneurial services—less production, more prosumer training—reduce dependence on monetary income and could be transferred to textiles, household appliances, furniture, tools, vehicles, food, etc.

This may seem like "paradoxical business economics", especially since companies could gradually make themselves dispensable through such prosumer management, because they release demanders into—mind you, partial—independence of consumption actions. But when the next resource or financial crises expose the modern fairy tale of perpetual growth in prosperity as a historical error, those companies will be competitive that help their customers to survive frugally but with dignity.

4.9 Conclusion

The sustainability discourse has ignored the findings of thermodynamics for too long. The well-known bon mot "there is no free lunch", often misused by neo-liberal protagonists, is confirmed in the material dimension. The works of Georgescu-Roegen [232] and Kümmel [2] support this insight. Thermodynamics points beyond physics; it supports a philosophy of the material zero-sum game and thus shakes the modernist belief in progress. From this derives not only economic rules based on limitation and moderation, but the realization that sustainable development cannot be an art of additional effect in the sense of green growth, but only an endeavor of creative omission and preservation.

By suppressing this—of course uncomfortable and politically not very opportune—fact, it was possible to pile up green promises of pleasure without regret, which may make the inclined public feel good, but correspond to a materially bounced cheque. The idea, still dominant in traditional economics, according to which (supposed) successes of industrial society, which are also aspired to by developing and emerging countries, would be based on increases in productivity and efficiency, the side effects of which are tolerable within the biosphere of our planet, must be rejected as a delusion. Consequently, it is time to finally accept the challenge posed by thermodynamics and translate it into an adequate economic design. This is exactly what is intended by the concept of post-growth economics presented here in excerpts. Sustainable development worthy of the name can only consist in promoting a reduction of aspirations and self-restraint.

5 Countries in Disruption

In industrial societies, energy and creativity accelerate technical progress and bring about upheavals. This also affects those countries that have found it more difficult to break away from agrarian modes of production and social structures than today's highly industrialized countries.

While Chaps. 1, 2, 3, 4 and 6 describe physical foundations and economic theories of economic activity in the past, present and future, the effects of value decisions are the subject of this fifth chapter.

Value decisions can have both positive and negative consequences. According to the model of economic growth outlined in Fig. 3.4, they are a component of the creativity factor. The economic slumps and recoveries after the two oil price shocks and the Lehman Brothers bankruptcy, which are important for the empirical testing of the model, were consequences of value decisions: for war and peace as well as for unrestrained speculation and economic policy reactions.

How the other two components of creativity, ideas and inventions, added new machines to the capital stock and improved its efficiency in the course of industrialization was described in Chaps. 1, 2, and 3. Appendix A.4 describes mathematically the final state of Industry 4.0, in which, thanks to the micro- and nanostructuring of the transistor invented by *Bardeen, Brattain* and *Shockley in* 1947, the production of goods and services would be totally automated according to the evolutionary principle of the factors of production formulated in Sect. 3.3.1. Moreover, we are witnessing how the advances in information processing made possible by the transistor have driven the development and expansion of the Internet and the miniaturisation of telecommunications through smartphones, and how these are swelling the flows of information around the globe and the flows of migration. The latter, and how value decisions can influence the quality and quantity of the factors capital, labour and energy, for example through changes in the structure of energy supply and the level of education in a country's resident population, will be addressed further on, as will political decisions and their economic consequences. In this chapter, which is partly subjective, I report personal experiences to the best of

R. Kümmel et al., *Energy, entropy, creativity*,
https://doi.org/10.1007/978-3-662-65778-2_5

my knowledge and belief. The same applies to information from the public media and the press.

First, let us look at Germany, whose rise from ruins to become the most economically powerful member of the European Union I have witnessed. It is now struggling with the conflict between economy and ecology and the growing gap between rich and poor, both nationally and internationally. Then we look to Colombia, to which I am attached through moving encounters with its people, the euphony of their language and the beauty of its nature. Its industrialization is stuck in feudalism, and the huge gap between rich and poor since colonial times has driven this country into the longest of all civil wars in Latin America.

I hope that in looking at these two very different countries from a point of view that does not ignore the sorrowful events in their history, the necessity of industrialization under well-considered reforms of their ever more complex progress will become empirically recognizable beyond all theory.[1]

There is a scientific reason for considering Colombia, in addition to the emotional one already mentioned. I would never have dealt with thermodynamics and economics if my Colombian colleagues in the Departamento de Física of the Universidad del Valle in Cali had not helped me to understand thermodynamics correctly for the first time. For they had assigned me the task of teaching this field, which until then had seemed to me to be boring and difficult to understand physically, as the first course in the master's program in physics that was to be established, and they had recommended F. Reif's excellent book *Fundamentals of statistical and thermal physics* [162] to me as a basis.

Having understood entropy and after the Club of Rome had published *The Limits to Growth* [95], it was clear to me that industrialization of all developing countries along the path followed so far by the industrialized countries will have an intolerable impact on the natural foundations of life on Earth. And the master's program in physics was, of course, intended as a contribution to laying the foundations for the industrialization of Colombia. Since then, I have been troubled by the limitations imposed on global economic development by the laws of thermodynamics.

This uneasiness has led me into the study of economics and its social problems. These have always included the causes of rich and poor and the social tensions that arise from them, up to and including armed conflicts. Since the 1970s, conflicts have also arisen between economic growth and environmental protection. Conflicts are complex. They cannot be understood without attention to the social and political facts associated with them, and their appreciation requires formulations that also make clear the emotional aspects of conflicts. This distinguishes this chapter in form and content from the other parts of the book.

[1]Perhaps in Colombian reforms, Niko Paech's "paradoxical business economics" of Sect. 4.8.7 could provide a new kind of livelihood for Colombia's youth and demobilized guerrilla fighters. And if one wanted to perform econometric analyses of economic growth even for developing countries like Colombia, the production function given in Eq. (A. 47) might be useful. However, the collection of reliable empirical data is likely to be more difficult than in highly industrialized countries.

5.1 Germany

The Thirty Years' War from 1618 to 1648 devastated Germany and left a politically unstable entity in the middle of Europe.

For a 100 years, with the disintegration of the religious unity of the "Holy Roman Empire of the German Nation" as a result of the Reformation that began in 1517, tensions between the German princes had grown in a disastrous concatenation of confessional and power-political interests. Occasional discharges in local military conflicts did not reduce them, on the contrary. The revolt of the Defenestration of Prague, small in itself, then triggered the great catastrophe in which the mercenary armies of the hostile German princes and the Spanish, French, Danes, and Swedes allied with them devastated central Europe. In the cultural and economic collapse of the Thirty Years' War, Germany lost about 40% of its population and national wealth. It disintegrated into more than 300 states, ruled by absolute monarchs and bishops who often lavishly and sometimes ridiculously tried to imitate the French "Sun King" Louis XIV. The German transition from a feudal agrarian economy to an industrial, bourgeois-capitalist market economy fell far behind that of England. Feudal structures survived in Germany until the early twentieth century. When the Germans, with their fractured national identity, finally combined their economic and scientific ventures organizationally in the German Customs Union of 1830 and in the Prussian-dominated empire founded in 1871, quite successfully, and caught up with their neighbors in industrial expansion, belatedly but powerfully, they exhibited the typical behavior of insecure laggards that prompted Winston Churchill (1874–1965) to remark, "The Germans are either at your feet or at your throat."

Having arrived in the twenty-first century as a passable democracy and almost normal country within the European Union, Germany is now trying to lead the way in climate protection and humanitarianism. The intention to do good in the world after the Nazi crimes certainly plays a role. But this creates technical and social problems that also affect the European Union. They are not being addressed for any party-political reasons, but because of the danger of Germany overreaching itself again in the current upheaval marked by thermodynamics and economics.

5.1.1 Energy Turnaround

> The decision to phase out nuclear energy here in Germany was the right one ....We now have the chance to show the world that a highly industrialised country is capable of converting its economy to a form of energy that is compatible with the preservation of nature. (*State Bishop Heinrich Bedford-Strohm, 2012* [163])

> The energy turnaround is nothing other than an operation on the open heart of the national economy. (*Federal Environment Minister Peter Altmaier, 2012* [164])

The perception and assessment of nuclear energy risks have changed considerably in Germany since the early 1970s.[2] At that time, with great popular approval, politicians had urged the reluctant, partly publicly owned energy utilities to use nuclear reactors to generate electrical energy. Again with great popular approval, the German government decided after 11 March 2011, in the wake of the Fukushima disaster and reversing the nuclear power plant lifetime extensions it had introduced in 2010, to shut down eight nuclear reactors immediately and the remaining nine German fission reactors forever by 2022. In 2012, the German Chancellor Angela Merkel described what Germany had set out to do with this in the words:

> An economic, environmentally friendly and reliable energy supply is one of the greatest challenges of the 21st century. For the German government, there is no question: we want to develop our country into one of the most efficient economies in the world while maintaining competitive energy prices and a high level of prosperity, and we want to reach the age of renewable energies faster than originally planned. [166]

Meeting this challenge is made more difficult by the fact that in the wealthy industrialized countries, and especially in Germany, the risks and side effects of an energy source are perceived all the more intensely and classified as unacceptable the more this source contributes to meeting energy needs. Nuclear energy is an example of this, and wind energy seems to be following suit in the wake of Germany's energy turnaround: numerous opponents of nuclear energy are now protesting vehemently against the "landscape-disfiguring" wind turbines and the power lines that are supposed to transport wind power from the windy north to the less windy south of Germany, where most of the remaining nuclear power plants are awaiting shutdown. Headlines were made on 2 January 2018 by a dpa report that grid operator Tennet had to pay almost one billion euros in 2017 for emergency interventions to stabilise the power grid due to a lack of lines for wind power. These costs are passed on to consumers via the electricity price. Apparently, neither the rising electricity costs matter to the protesters, nor the fact that wind power plants are burdened with the lowest life cycle CO_2 emissions of all renewable energies, as Table 1.2 shows, and that according to Table 2.6 they contributed the most to electricity generation from renewable energies in 2016 with 12.1%.

Fukushima

One of the most severe earthquakes in the history of Japan and the resulting tsunami with wave heights between 13 and 15 m destroyed four reactor units of the Fukushima 1 nuclear power plant (NPP) on 11 March 2011. In units 1–3, the failure of the cooling systems caused the metallic fuel elements in the reactor core to melt (core meltdown). Reactor 4 was shut down for maintenance. Its fuel elements, which generate a lot of residual heat, were stored in the decay pool inside the reactor building. Its cooling system also failed. The water in the decay pool heated up. On

[2] The passages up to and including the *risk of a GAU in German nuclear power plants are* based on [165].

15 March 2011, Unit 4 exploded, probably due to hydrogen release and ignition of the with oxygen generated oxyhydrogen gas. In total, the destruction of units 1–4 caused about 10–20% of the radioactive emissions of the Chernobyl accident.

The power plant units of Fukushima 1 were built between 1970 and 1978 on the Japanese coastal part of the Pacific Ring of Fire according to design plans for boiling water reactors that the company General Electric had developed primarily for the NPPs that had gone into operation in the USA in the 1960s. Accidents in three Japanese nuclear power plants caused by earthquakes in 2005 and 2007 showed that the design of the reactors, especially those built in the 1970s, did not take into account the possible earthquake magnitudes in Japan. But this was accepted. The same applies to the tsunami risks. Japanese classical paintings as well as films from the twentieth century show tsunami waves rising higher than 20 m. However, for the seaward part of the Fukushima-1 site, only a 5.70 m high protective wall existed, and only 3.12 m was prescribed. Reactor units 1–4, located 10 m above sea level, were flooded to a depth of up to 5 m. The emergency generators in the basement of the turbine buildings were only a few metres above sea level and were inadequately protected against flooding. Thus, the earthquake destroyed the connection of the reactor units to the power grid, and the tsunami paralysed the emergency generators. Despite the rapid shutdown of units 1 to 3, the residual heat of the fuel elements could no longer be dissipated due to a lack of cooling, and the disaster occurred [167, 168].

Residual Risk

After the series of accidents in Fukushima 1, the German government purported the existence of an underestimated residual risk of nuclear energy as justification for withdrawing the previously agreed NPP lifetime extensions and the decision to completely phase out nuclear fission energy. Before the 2009 Bundestag elections, the CDU/CSU and FDP left no doubt about their intention to replace the Red-Green nuclear consensus of 2002, which had provided for a time limit of 32 years for the normal operating lives of NPPs since commissioning, with the operating life extensions.[3] Voters then gave them a comfortable majority in the Bundestag. And, a comparably large majority of the German population then probably also went along with the abrupt turnaround in nuclear energy policy after March 2011. Whatever may have prompted this majority to do so, it is irrational to draw conclusions from Fukushima 1 about hitherto unknown and widely underestimated residual risks of German nuclear power plants. After all, the residual risk describes the dangers of a system despite existing safety systems. It consists of an assessable and an unknown part. For example, the simultaneous, accidental failure of safety systems can be estimated. Unknown, on the other hand, is the probability of terrorist attacks. But it was not the realisation of a residual risk that was responsible for the Fukushima disaster, but an incorrect, inadequate design of the plant against

[3]The lifetime extensions were 8 years for the seven plants built before 1980 and 14 years for the remaining ten nuclear reactors.

earthquakes and tsunamis, in other words the deliberate acceptance of a known risk. In Germany, on the other hand, the Mühlheim-Kärlich NPP, built in the Rhine valley near Koblenz and commissioned on 1 March 1986, was taken off the grid on 9 September 1988 due to procedural errors in construction law in connection with geological risks; dismantling began in the summer of 2004.

Risks of Water- and Graphite-Moderated Nuclear Reactors

The immediate risk of nuclear energy lies in the possibility of an accident in which all cooling systems fail and large quantities of radioactive material are ejected. The moderator plays a central role in the risk analysis. The fast neutrons released in a nuclear fission cannot split any other atomic nuclei. They must first be slowed down to thermal speeds by the moderator in order to be able to maintain the chain reaction. The boiling and pressurized water reactors of western design use water as moderator. The RBMK reactors developed in the Soviet Union use graphite.

Harrisburg

In water-moderated nuclear reactors, the water also serves as a coolant from which the steam is produced that drives the electricity-generating steam turbines. The cooling water circuit is maintained by pumps and ensures the correct operating temperature of the fuel elements in the reactor core. If an accident causes the cooling water circuit to stop, the fuel elements heat up and steam bubbles form in the boiling water. Since the water density in the steam bubbles is low, neutron moderation is severely weakened. This is referred to as a negative temperature or bubble coefficient. In other words, insufficiently fast neutrons are slowed down to thermal speeds by collisions with water molecules, and the chain reaction comes to a standstill. The reactor is shut down even faster by inserting neutron-eating control rods. Nevertheless, in the worst case, the residual heat that the fuel elements continue to generate due to beta decay for hours or days after nuclear fission has ceased can lead to the melting of the metallic fuel elements in the reactor core if cooling is inadequate. Such an accident occurred on 28 March 1979 at the Three Miles Island NPP southeast of Harrisburg in the US state of Pennsylvania. After a failure of the cooling water pumps, the rapid shutdown of the pressurised water reactor worked. However, technical defects and operating errors caused a partial meltdown of the reactor core and the release of radioactivity. The plant was severely damaged. This accident caused opposition to the peaceful use of nuclear energy to grow strongly in the western industrial nations.

Chernobyl

In RBMK reactors, the rod-shaped fuel elements are embedded in moderating graphite. They are cooled with water. In contrast to water-moderated reactors, the temperature coefficient (bubble coefficient) is positive here. The reason for this is that the cooling water also always absorbs a certain proportion of the neutrons and that this must be taken into account when designing the reactor for normal operation. Now, if for some reason the reactor cooling system fails, the graphite and water heat up. Steam bubbles form in the boiling water. In this case, because of the low water density in the steam bubbles, neutron absorption is significantly reduced while moderation by graphite continues. This dramatically increases the production of thermal neutrons. If control rods are not inserted to reduce neutrons, more and more thermal neutrons accelerate the chain reaction and the reactor explodes. Such an accident occurred on 26 April 1986 in reactor 4 of the Chernobyl NPP near Kiev, in a poorly organised safety experiment in which an inadequately trained operating team deliberately disregarded a number of safety regulations and made many operating errors. This, in conjunction with the positive temperature coefficient, led to a

> power excursion of the nuclear fission chain reaction by a factor of about 500, this within less than a minute for the duration of about a minute. The resulting explosion terminated the power excursion, destroyed the reactor pressure vessel, destroyed the reactor building (which was not constructed as a pressure-protected enclosed building), set fire to the graphite moderator, resulted in a release of all volatile radioactive materials and about 4 percent of the total inventory of radioactivity in the reactor. [169, p. 257]

Risk of a GAU in German Nuclear Power Plants

A greatest imaginable accident (GIA, in German GAU) releases a large part of the radioactive inventory of a NPP. An accident like the one in Chernobyl cannot occur in German water-moderated reactors because of their negative temperature coefficient. An accident like in Fukushima 1 is as likely as the destruction of the emergency generators of German NPPs by a tsunami. Nevertheless, due to the production of residual heat after the chain reaction has gone out, a GAU cannot be ruled out 100% even in the existing German nuclear power plants. The residual risk is estimated at 1 GAU per million years of reactor operation; in Germany, therefore, a GAU could have occurred once within 50,000 years during the operation of the 20 reactors that produced electricity in the mid-1990s [169, p. 227]. The first problem of an ethical evaluation of nuclear energy is therefore: Is an energy technology acceptable for which an accident with very large consequential damage can occur with a very low probability? The radioactive emissions, the damage to health, the deaths and the land contamination as a result of the Chernobyl accident are an example of such consequential damage [169, p. 255], which cannot be ruled out with certainty even with the very low residual risk of the German NPPs still in operation in 2011.[4] This contrasts with the known damage and risks to health and the environment from the use of fossil fuels with which we have been familiar for 200 years. What is still unknown is the full extent of the adaptation pressure on a growing earth population that climate changes generate, which exceed all climate fluctuations during the history of civilization of mankind.

[4]In the high-temperature reactors developed at Forschungszentrum Jülich, one of which, the thorium high-temperature reactor THTR-300, went on line in Hamm-Uentrop in 1983 and was shut down for good on September 1, 1989, mainly for economic reasons, a core meltdown is not possible for physical reasons; more on this in [2, pp. 79–81].

CO_2 Emissions and Costs of German Electricity Generation Since the Proclamation of the Energy Turnaround

After the Harrisburg accident, the first proclamation of a German "energy turnaround" was made in 1980, by a study of the Öko-Institut Freiburg. The eco-classic *Energiewende—Wachstum und Wohlstand ohne Erdöl und Uran* [170] proposed an alternative economic growth in the sense of the programme mentioned in the title. It spoke of the citizens who do not want Germany to become a "Harrisburgland" and coined the term "dinosaur technology" for the civilian use of nuclear energy. This term is readily used by the anti-nuclear movement as an expression of its contempt.[5]

The public debate revolved around the use of renewable energies, increasing energy efficiency and transforming the German economy into a "blueprint economy". Its prosperity would be based on the production of patents and ideas. These would be exported in the form of licenses, patterns for the textile industry, planning services for buildings and transport infrastructure, organizational schemes, construction plans for industrial plants, and more. We would import the energy-intensive products needed in Germany and pay for them with the proceeds of our high-priced intellectual creations. The fact that at that time the global production of goods and services relied at least as much as today on the burning of fossil fuels, see Fig. 2.1, and German demand for energy-intensive imported goods contributed to CO_2 emissions, was not an issue. Although the Swedish physical chemist Svante-Arrhenius had already pointed out the anthropogenic greenhouse effect in 1895, this was not seen as a problem until the mid-1980s [12].

The energy turnaround of 2011, falsely justified by the "residual risk", was introduced into law—after fierce protests against the extension of nuclear power plant lifetimes, shortly before an important state election, and more far-reaching than the "energy turnaround" called for by the publication "Energiewende" in 1980. Even if this was primarily about abruptly phasing out nuclear energy, the "energy turnaround" is now understood by those with political responsibility and large sections of the public as a new climate protection policy. Germany wants to show the world how an energy supply structure based on centralized energy conversion plants and fossil and nuclear energy sources that store solar and cosmic energies can be transformed within a short period of time, without any system-analytical preparation, into a decentralized energy supply structure that depends on renewable energies that fluctuate statistically with the wind and weather.

These arise from the solar energy flows to the Earth, which, as already mentioned in connection with Fig. 2.1, are around 10,000 times greater than the world's energy demand in 2012. The entropy production associated with their generation in the sun

[5]This apparently does not take into account that the dinosaurs were the most successful land vertebrates to date and dominated the earth for 170 million years before suddenly disappearing about 65 million years ago. The cause of their mass extinction was lack of solar energy and biomass due to a meteor impact and/or violent volcanic eruptions. The large masses of dust thrown into the atmosphere reduced solar radiation, and the lush vegetation, the dinosaurs' food base, withered away. How our approximately 2-million-year-old species Homo, and in particular a Homo sapiens solely dependent on solar energy, would survive such a catastrophe, we do not know.

is lost in the vastness of space and does not burden the biosphere of our planet. These are the advantages of renewable energies from the sun and wind that cannot be surpassed in the long run.

The problem is the transition to their large-scale industrial use. Because this transition costs money. Even in a relatively wealthy economy like Germany's, with a national debt of EUR 2000 billion at the end of 2016, equivalent to 68.3% of GDP (France 96%, UK 89.3%, Italy 132.6%), every effort should be made to make the transition cost-minimizing. The tools to optimise energy, emissions and costs are available in the major energy research institutes in our country and the European Union. And as discussed in Sect. 2.3.3, even optimization models limited to small regions can provide useful guidance.

But Germany's energy turnaround has not followed this path, but rather that of a Heave Ho!-process. Moreover, it has not taken into account the well-known NIMBY (Not In My BackYard) mentality that prevails among broad segments of the population in wealthy industrialized countries and that often opposes new infrastructure projects.

Let us compare the goals of the energy turnaround with what has been achieved so far.

Targets

The emission reduction targets from 2010 were continued unchanged after 11 March 2011 until 2050, despite the fact that nuclear power plants with a share of 22.2% of German gross electricity generation in 2010 and average CO_2 life-cycle emissions of 10–30 grams per kilowatt hour (g/kWh) must be completely replaced within 12 years by renewable energies with a share of 15.8% of German gross electricity generation in 2010 and average CO_2 life-cycle emissions of between 10 and 20 g/kWh (wind) and 70–150 g/kWh (photovoltaics) (Tables 1.2, 2.5, and 2.6). The production principle mentioned in Sect. 2.3.3 is used for the emission quantities and not the more appropriate consumption principle compared with it in Table 2.4. According to the latter, German CO_2 emissions in 2011 would have to be multiplied by a factor of 1.4. The more energy-intensive production is relocated from Germany to other countries, the greater this factor becomes.

According to the German government's energy transition plans, the share of renewable energies is to increase, namely to 36% of electricity generation by 2020 and to more than 50% of primary energy consumption by 2050. And German greenhouse gas emissions relative to emissions in 1990 are to fall, namely by 40% by 2020, 55% by 2030, 70% by 2040 and 80% by 2050, assuming an annual increase in "energy efficiency", understood as GDP/(annual primary energy consumption), of 2.3–2.5%. Furthermore, one million electric cars are to be on Germany's roads by 2020.

By the end of **2016, the** following had been achieved:

1. **Energy**

According to Table 2.6, the share of all renewable energies in *gross electricity generation* is 29%. Of this, 6.9% is accounted for by biomass and 0.9% by household waste.

The publication *Die Energiewende: unsere Erfolgsgeschichte (The energy transition: our success story) by* the German Federal Ministry for Economic Affairs and Energy [171] states that renewable energies accounted for 32.3% of *gross electricity consumption* in Germany in 2016. One explanation for the discrepancy with the 29% may be that the ministry only includes renewable energies in domestic electricity consumption and not in gross electricity generation, a growing proportion of which has been exported since 2003, see Table 5.2 and related explanations.

According to Table 2.7, the shares of renewable energies in *primary energy consumption* for photovoltaics, wind and water (PWW) total 3.9% and for fuels (biomass, household waste: BEE) 8.7%.

Two assessments of the biomass:

> It does exist: the energy transition master plan. It is the final report of a research project commissioned by the Federal Environment Ministry. It comes from the German Aerospace Center in cooperation with other institutes .... The study comes to the following conclusions: Biomass will continue to account for the vast majority of renewable energies in the future. In 2030 it will still have a 46% share of renewable energies. [172]

The German National Academy of Sciences Leopoldina [174] points out that.

> photovoltaics, solar thermal energy and wind turbines usually have an efficiency per unit area (W per square metre) ten times higher than plant photosynthesis

and affirmed that

> the use of biomass falls sharply with an EROI[6] [173] of mostly less than 3. Of the alternative energy technologies, biomass-derived energy contributes the least to reducing GHG emissions[7] and has the highest financial cost per tonne of CO_2 saved.

With regard to the

> ecological risks, climate and environmental costs…an expansion of land for the cultivation of energy crops seems ecologically questionable. It is likely to contradict existing regulations on the protection of biodiversity and nature … .In life cycle analyses of biofuel production and consumption, … the following additional environmental costs must be taken into account: Changes in soil quality and biodiversity; contamination of groundwater, rivers and lakes with nitrates and phosphates; and in the case of irrigation, negative effects on groundwater levels and soil salinisation.

---

[6] Energy Return On (Energy) Investment.

[7] Greenhouse gas emissions.

Table 5.1 German emissions of greenhouse gases (GHG, in carbon dioxide equivalents) and CO_2 [175]

Year	1990	2000	2003	2010	2013	2015
GHG (million t)	1251	1070	1050	930	950	910
CO_2 (million t)	1031	905	915	834	835	800

Despite its environmental problems, biomass plays a key role in German climate protection concepts, both at federal and municipal level, because, like fossil and nuclear energy sources, it is an *energy store.*

The batteries of the 1,000,000 electric cars envisaged by the German government for 2020 could also serve as storage for electric energy, provided that the infrastructure for charging stations, electricity transport and user behaviour develops appropriately. It is not yet certain that urban grids will be able to cope with the strain of charging a rapidly growing number of electric cars. Premiums encourage the purchase of electric cars. According to the Federal Motor Transport Authority's announcement of 2 March 2017, on 1 January 2017 there were around 34,000 pure electric cars on the road (+34% relative to the previous year) and around 165,400 (+26.8%) hybrid cars out of a total of around 45,804,000 passenger cars.

Electromobility pioneer Tesla has yet to make a profit in the 8 years leading up to 2017. It is also uncertain whether the materials required for the mass production of electric cars and the associated infrastructure can be purchased on the world market in the future at the previous prices. Possible material bottlenecks are described in [93].

2. **Emissions**

The development of German greenhouse gas and CO_2 emissions since 1990 is shown in Table 5.1.

The base year 1990 agreed in the international climate protection targets and agreements is favourable for Germany in the comparison of coutries concerning emission reductions. Since this year of German reunification, the production plants of the former GDR, whose energy efficiencies were significantly lower than those of the old FRG, have been shut down.[8] This greatly benefits the emissions balances of the new FRG.

Added to this is the trend towards relocating both personnel-intensive and energy-intensive firms abroad and the intensifying car and truck traffic on German roads in the course of globalisation. Both of these factors have an impact on the German emissions balance. The effects of the energy turnaround should be most clearly

[8] "The industrial production here in the East is easily taken over by your factories in the West," said to me in the spring of 1991 the manager of an iron foundry in Guben, who was married to one of my wife's (pen-pal) friends, pointing to the date 1876 above the entrance to the brick factory hall that he would soon have to close. "The last person I'm going to fire is myself." And so it came to pass soon after.

Table 5.2 Electricity consumption and CO_2 emissions in Germany [176]

Year	1990	2000	2003	2010	2013	2015
I (million tonnes)	366	327	340	316	331	309
II (TWh)	482	510	538	565	570	578
III (g/kWh)	761	640	633	558	580	534
IV (TWh)	482	514	530	548	537	527
V (g/kWh)	759	536	643	570	617	587

I CO_2 emissions from electricity generation, *II* electricity consumption, *III* CO_2 emissions factor for the electricity mix, *IV* domestic electricity consumption, *V* CO_2 emissions factor for domestic electricity consumption—"electricity consumption" means consumption of electrical energy

visible in the specific emissions of German electricity generation, which are shown in Table 5.2.

Explanatory notes to Table 5.2:

1. Electricity consumption = gross electricity generation minus power plant own consumption minus pumped-storage line losses.
2. Domestic electricity consumption = Electricity consumption including electricity trade balance = gross electricity generation minus power plant own consumption minus pumped-storage line losses *plus* electricity imports minus electricity exports.

From 2003 onwards, the electricity trade balance, i.e. electricity exports minus electricity imports, is positive. In 2015, it amounted to 51 TWh (difference between row II and row IV in Table 5.2).

Conclusion from Table 5.2:

Specific CO_2 emissions per kWh of electricity consumed domestically (row V) fell from 759 to **570** g between 1990 and 2010. They then increased to 617 g in 2013 and decreased to **587** g in 2015. Emissions related to domestic consumption have *increased* by 3% compared to 2010 as a result of the energy turnaround, which has so far mainly affected the electricity industry.

3. **Costs**

The Renewable Energy Sources Act (EEG) regulates the preferential feed-in of electricity from renewable energy sources into the German electricity grid and contractually guarantees producers fixed feed-in tariffs. Its predecessor was the Electricity Feed-in Act of 1991. The initial version of the EEG in 2000 was followed by amendments in 2004, 2009, 2012 (entry into force on 1 January 2012, amendments at the end of June 2012 by photovoltaic amendment), 2014 and 2016/2017. The many amendments are due to the cost issues associated with the EEG. As an example, Fig. 5.1 shows the development of the photovoltaic generation capacity and the total feed-in tariff paid annually by electricity customers, and the expected future development before the 2016/2017 amendment as well. According to the actual development reported by [178], there was an increase in generation

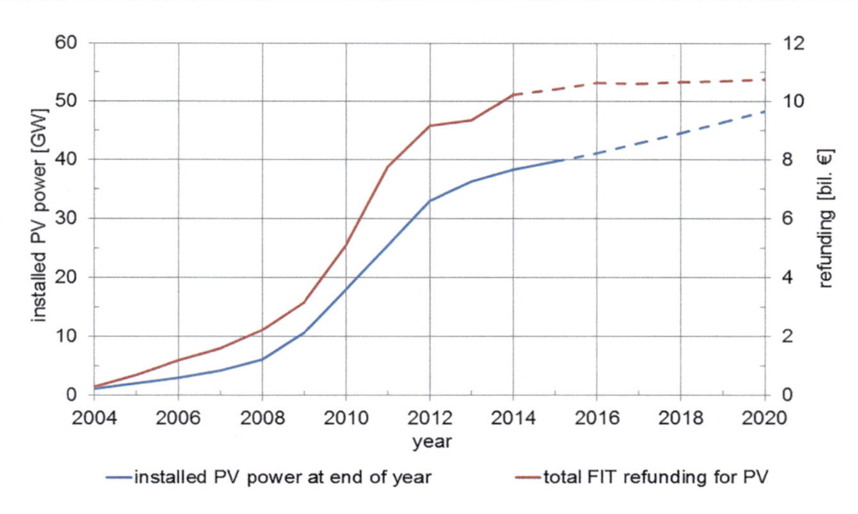

Fig. 5.1 Development of installed photovoltaic generation capacity (in gigawatts, left ordinate, lower curve) and feed-in tariffs according to the EEG (in billions of euros, right ordinate, upper curve) in Germany. Solid curves: empirical data, dashed curves: projections. (This figure was taken from [96] and reproduced there with kind permission of H. Wirth [177]. Current data and PV expansion strategies are presented [178])

capacity to 41 GW by 2016, a feed-in tariff of about EUR 10.6 billion in 2015 and a decrease of the same to EUR 10.1 billion in 2016.

In its success story of the energy transition [171], the Federal Ministry for Economic Affairs and Energy reports as part of the success: (a) on p. 6 the reduction of the average subsidy level for large photovoltaic systems from 9.17 ct/kWh in April 2015 to 6.9 ct/kWh in December 2016 and (b) on p. 11, that the cost dynamics of the electricity price had broken: the average electricity price for a household with an annual consumption of 3500 kWh rose from 23.96 ct/kWh in 2010 to 29.14 ct/kWh in 2014; then it declined to 28.80 ct/kWh in 2016. In 2006, it had been 19.46 ct/kWh.

Before the Fukushima catastrophe, the fuel element tax was introduced in 2011 and retained after the phase-out of nuclear energy. The electricity companies filed a lawsuit against this. In June 2017, the Federal Constitutional Court ruled that the fuel element tax was unconstitutional because it was not a consumption tax but a tax on a means of production. The state was ordered to repay the withheld EUR 6.3 billion with interest to the suing electricity companies E.On, RWE and EnBW.

Another of the cost problems of the forced restructuring of the German energy system is that on sunny and/or windy days the wanderer in German lands sees many idle wind turbines whose owners have to be compensated for the curtailments according to the provisions of the Renewable Energy Sources Act (EEG), and that nevertheless still more "green" electricity enters the grids than is demanded in Germany. This surplus electricity is then exported, mostly to Austria and increasingly at negative prices, e.g. on Sundays, and not infrequently re-imported shortly afterwards, e.g. on Mondays, as green electricity from Austrian pumped storage power plants at peak prices.

New, highly efficient and relatively low-emission gas-fired power plants, which can be quickly ramped up and down and react flexibly to fluctuations in demand on the one hand and the fluctuating feed-in of wind and solar power on the other, became uneconomical after 2011 and were partly shut down because their relatively expensive electricity could no longer be sold at a profit on the Leipzig power exchange during the midday demand peaks, but was displaced by the "right-of-way", feed-in tariffed green electricity. Whether the situation will be improved by the latest EEG amendments with more competitive elements remains to be seen. In any case, in November 2017, the Siemens Group announced global plans to cut 6900 jobs in its power plant and propulsion divisions, partly because of the collapse in demand for large gas turbines.

The displacement of gas-fired power plants from the electricity market exacerbates the already existing problems of grid stabilisation.

> The regenerative energy feeds do not contribute to the power frequency control. This is because they run via a DC link and a grid frequency controlled inverter and are therefore dynamically decoupled. Grid operators have to intervene several times a day to ensure grid stability. In February 2012, a grid collapse caused by the failure of sun and wind, a so-called blackout, could only just be prevented. [179]

Our partners in the European electricity network are no longer idly watching Germany forcing its wind power from the north through the Polish and Czech grids, destabilising them in the process of exporting electricity to Austria, which in some cases involves negative prices, due to the lack of sufficient north-south high-voltage lines of its own. This was stated in a press release from the energy industry on 20 December 2016:

> **Germany-Austria electricity price zone split from 2018: more negative prices, more EEG §51 losses.**
>
> If negative electricity prices occur for at least 6 hours at a time, EEG plants do not receive a market premium for these periods. This regulation was already found in the EEG 2014 for plants ≥500 kW and wind turbines (WT) ≥3 MW with commissioning from 2016. The EEG 2017 maintains this, but with the Amendment Act to the EEG of 16.12.2016, the (re-) introduction of the bracketing was additionally decided, which leads to the fact that §51 de facto also applies to WT ≤3 MW in a wind farm.
>
> However, EEG investors are particularly concerned about the planned division of the electricity price zone between Germany and Austria. This will inevitably result in more negative prices in the future and thus more §51 losses. The decision now taken by the Federal Network Agency to dissolve the uniform electricity price zone and to introduce congestion management is based on an initiative by the European regulatory association ACER. ACER had called on Germany and Austria to break up the uniform price zone, which has existed since 2001, in order to address increasing grid problems in Poland and the Czech Republic caused by German electricity exports. The BNetzA intends to implement this separation in winter 2018/2019.
>
> In terms of energy economics, the introduction of congestion management means limiting the flow of electricity to Austria. The congestion then comes into effect above all in times of high RE generation in Germany. If low demand and high and cheap feed-in

> coincide and export capacities are exhausted at the same time, negative electricity prices often arise. Scenario calculations with the enervis electricity market model confirm that the frequency and duration of the §51 case will increase in the future due to the reduced grid coupling capacity at the German southern border. A flexibilisation of the power plant fleet and the electricity market-driven use of storage facilities can dampen this increase proportionally—however, compared to the previous situation of unlimited electricity trading with the Alpine Republic, more negative prices and more §51 losses are definitely to be expected. [180].[9]

According to a dpa report on 4 January 2018, the number of hours during which German electricity had to be exported at negative prices rose from 15 in 2008 to 146 in 2017.

The cost problems that have arisen with the energy turnaround since 2011 can obviously no longer be ignored by politics, business and private households. In terms of ethics, one might dismiss this with: "How can one be pettyfying about cost, if the world has to be saved from climate catastrophe!"

For this purpose, it is worth taking a look at the percentage shares of the six strongest CO_2 emitters in global CO_2 emissions in 2011: (1) PR China 26.4; (2) USA 17.7; (3) India 5.3; (4) Russia 4.9; (5) Japan 3.8; (6) Germany 2.4. [181] (Between 2010 and 2015, German CO_2 emissions still decreased somewhat, according to Table 5.1 from 834 to 800 million t/year).

If Germany were to ban all internal combustion engines, shut down all fossil fuel power plants, bring the housing industry up to passive house standards and make the manufacturing sector CO_2-free, and if all other countries were to emit as much as they did in 2011, global CO_2 emissions would fall by no more than 2.4%.

Germany's contribution to a significant reduction in global CO_2 emissions can only consist of our country taking a technically and economically exemplary path towards this goal, so that other countries are also willing and able to follow this path.

Even if German climate protection activists still see the Federal Republic in a pioneering role and "Energiewende" has found its way into other languages as a loanword—the world now sees "climate protection" differently from Germany. In 2017, Swiss citizens decided in their referendum not to build any more new nuclear power plants, but to keep the existing ones running as long as they meet safety requirements. The United Nations recommends replacing fossil fuels with renewables and nuclear energy to reduce CO_2 emissions. And, "Globally, 79 are newly planned in 7 countries with a net capacity of 88,201 MW" [182].

In particular, Germany must not lose sight of its European partners. This is because the multiple interconnections and dependencies of the German electricity industry within the European electricity network mean that the

[9] Enervis energy advisors GmbH is a management consultancy for electricity and gas suppliers as well as investors and operators of renewable and conventional power plants and storage facilities.

> national climate protection plan … is not compatible with climate protection efforts at European level. Additional national measures of CO_2 regulation within the internal electricity market and European emissions trading system would only lead to a shift of electricity generation and emissions to our neighbouring countries. [183]

At the Bonn World Climate Conference in November 2017, EU members France, the UK, Italy and the Netherlands joined Canada and other countries to form the Powering Past Coal Alliance with the aim of phasing out coal. France met about 75% of its electricity needs from nuclear power in 2017. President Macron has promised to bring that contribution down to 50% by 2025 if possible. The UK's ageing coal-fired power stations have become uneconomic. The Hinkley Point nuclear power station on the southwest coast of England is to be expanded, also with Chinese participation, by two more nuclear reactors with a total of 3260 MW. Underneath Italy lie very large quantities of oil and natural gas; in 2014 it met only about 10% of its fossil fuel needs from them. Similarly, the Netherlands has large reserves of natural gas. Canada is the world's fifth-largest energy producer; more than 60% of its oil production comes from the Alberta oil sands. In 2013, it generated about 620 TWh of electricity, roughly the same amount as Germany according to Table 2.5, from hydropower—where Canada is the world's third-largest producer—coal, nuclear power and wind power [184]. After the last nuclear power plant is shut down in 2022, Germany will only have lignite available to operate the (still) indispensable base-load power plants without importing fossil fuels.

The Storage Problem and New Technologies

In the current discussions about the energy transition, energy storage systems, especially electricity storage systems, play a central role. However, the use of energy storage systems is making only slow progress in the German economy.

Pumped storage power plants, which store electrical energy with an efficiency of up to 75%, are affected in their economic viability by the competition from solar and wind power at peak times, similar to gas-fired power plants. In addition, the supply of pumped electricity is burdened with grid usage fees. The economic incentives for grid stabilization through German pumped storage power plants are low, if not non-existent. In North Rhine-Westphalia alone, there are 23 sites with a total technical and ecological storage potential of 56 GWh [185].

Studies on the use of large heat and power storage systems in combination with cogeneration showed [186]:

> The application of novel storage concepts, such as seasonal heat storage or the use of superconducting magnetic energy storage systems, can noticeably increase the primary energy savings that can be achieved in an energy supply system through combined heat and power (CHP), if the charging and discharging of the storage systems is optimally controlled. In addition to the technical characteristics of the CHP plants, it is above all the structure of the temporally fluctuating demand for heat and electrical energy that influences the primary energy savings potential. If the demand for heat and electrical energy is known with time resolution, even simple model calculations allow the primary energy savings that can be realised through CHP and storage to be determined. More precise results can be obtained by using dynamic optimization models.

"Simple model calculations" are those that assume an energy storage system called "ideal" that is capable of storing arbitrarily large amounts of heat and electrical energy without energy or exergy loss. The more precise results follow from the dynamic optimization of the charging and discharging management of real, lossy energy storage systems.

Although the seasonal heat storage and superconducting storage of electrical energy considered in the optimisation scenarios of [186] were already known in the 1990s, they still do not play a role in the German energy system in the second decade of the twenty-first century. This is because storage hardware is expensive. The same applies to the development and nationwide implementation of the software that optimizes their charging and discharging, which must carefully take into account local and seasonal conditions.

The DBI—Gastechnologisches Institut gGmbH Freiberg [187] writes on the storage of electrical energy via "power-to-gas"/methanisation:

> The storage of large amounts of electricity from fluctuating energy sources (wind, PV) by converting electrical energy via electrolysis into hydrogen has become known as power-to-gas. Since the natural gas grid can only absorb hydrogen to a limited extent, the option of methanating carbon dioxide with the hydrogen formed in electrolysis is an interesting option in the longer term. Methane can be fed into the natural gas grid in practically unlimited quantities. However, the additional process step of methanation is currently still associated with relatively high costs and efficiency losses.

Recent research on energy storage includes nickel nanocluster catalysts for the production of chemical fuels from renewable resources [188]. Artificial photosynthesis using semiconductor nanotechnology could produce synthetic gases and/or hydrogen from solar energy to serve as energy storage and fuel [189].

Biophysics, physical chemistry and systems analysis are joining forces with energy technology to open up paths to a sustainable energy economy. But the new processes should develop evolutionarily out of the existing energy system. It is enough that once a country—China—has had experience with a "Great Leap Forward".[10]

In order for the German energy turnaround to find its way out of the state of trial and error, German energy and environmental policy should turn away from banning the energy technologies that currently supply our country and turn towards promoting new, forward-looking methods of energy generation, distribution and storage, in cooperation with our European neighbours. In an appropriately restructured tax system, the innovations will prevail against the old, fossil energy system without bans on the market.

[10]The "Great Leap Forward" is the name given to a campaign that ran from 1958 to 1961 in the People's Republic of China, in which the ideological re-education of the population, neglect of agriculture and a shift towards decentralised industrial production in the countryside were intended to catch up with the western industrialised countries in a short space of time. It failed in a famine catastrophe.

5.1.2 Migration

> The myth of Germany makes her furious. (*Fiona Ehlers on Zineb Essabar, Italian-based black African woman who assists African refugees at the Brenner Pass* [190])

In physical and social systems, gradients lead to flows. In physical systems, temperature gradients and gradients of chemical potentials drive the heat and particle flows described in Eq. (A.4). In animal populations, geographic differences in food supply force long-distance migrations across lands, seas, and air. Additional migratory pressures arise in human societies from armed conflicts and information about regional and global wealth disparities.

Thanks to the transistor, people have known since the 1960s that hardship and poverty are not an inevitable fate of the less fortunate. "The, transistor," that is, the lightweight, battery-powered radio equipped with transistors, and its advertising messages inform Colombia's rural poor about the wonderful things available in the big cities, especially Bogotá, Cali, and Medellin. "That is why they flee from rural poverty and violence to the cities, swelling the slums there," reported the country expert for Colombia during a preparatory course by the Working Group for Development Aid.

On the social transistor effect in Colombia, an example: On a visit to a mountain village high in the southwestern Colombian Andes, I was thrilled by the view across the valley of the Patia River to the mountain ranges towering one behind the other in the distance, their lower regions tropical green and their peaks shining rock-red in the sinking sun under a deep blue sky. "How much better than the people in the barrios populares[11] of Cali you live here in this magnificent nature. Stay in El Rosario. Don't move to the city," I said, talking to the villagers. "Sí, senor," one replied, pointing to his transistor radio, from which blared the advertising slogans for upper-class luxury consumption, "but there are so many beautiful things in the cities – and what do we have here? It's just enough to fill us up. There's nothing else going on here." Even the references to unemployment, cramped quarters, stinking sewage and violent crime in the overcrowded slums of the big cities probably did nothing to change his dream of a better life in the city with its industrially manufactured consumer goods. That was then, in the 1970s, a relatively peaceful time in Colombia between the civil war of the "Violencia" and the drug and land theft wars that developed from the 1980s onwards and turned about 10% of the Colombian population into internally displaced persons; more on this in Sect. 5.2.

In the world population, which at the end of 2016 comprised more than 7.47 billion people, 60% in Asia, 16% in Africa, 9% in Latin America, 10% in Europe, 5% in North America, 0.3% in Australia, with Africa showing the strongest population growth at 2.5% [191], the wealth gap between North America, Europe and Australia on the one hand, and the industrially less developed countries on the other,

[11] Colombian euphemism for slum.

is driving migration flows, the temporal development and sources of which, using Germany as an example, are shown in Fig. 5.2 and Table 5.3.

According to Fig. 5.2, the number of asylum applications filed annually in the Federal Republic of Germany (FRG) since 1953 reached an initial maximum of around 438,000 applications in 1992. From 2013 onwards, it again skyrockets: from about 203,000 in 2014 to 477,000 in 2015 to 746,000 in 2016; from January to September 2017, about 168,000 applications were filed. That number rose to 187,000 by the end of 2017.[12]

One cause of flight and migration to Germany are the regional wars since the end of the Cold War. Another might be expectations of an economically strong Germany, the vast majority of whose citizens consider themselves lucky that their country has been accepted back into the community of nations after the crimes of the Nazi era, and even reunited. Asylum seekers are not deterred by the fact that in times of high immigration there are arson attacks on refugee shelters. Let us look at the migration impulses that preceded each of the two immigration peaks.

After the fall of the Berlin Wall and the Iron Curtain in 1990, the multi-ethnic state of Yugoslavia, internationally influential on its third way between East and West, North and South, disintegrated during the last decade of the twentieth century in a series of wars between its different ethnic groups. The Bosnian war with its ethnic cleansing raged the worst from 1992 to 1995. At that time, many war refugees sought refuge in the newly reunited Germany. This is one of the reasons for the first peak in asylum applications in 1992.

From 1993 until their repatriation to Sarajevo in 1997, we had housed two Bosnian Muslim refugee families, one temporarily rent-free, in two condominiums. We had a few experiences in the process, such as the following.[13]

State and church institutions in Germany may urge the population to provide living space for refugees. But if one asks those responsible for the allocation of apartments owned by these institutions whether demonstrably free living space can be rented to Muslim refugee families, after various false statements and excuses comes the rejection: "We don't want Muslims".

After the husband and father of the first refugee family, consisting of mother, son and daughter, who had stayed behind in Sarajevo, had fallen, the 9-year-old son became the head of the family according to Muslim understanding and demanded that his mother take on even more cleaning work so that she could buy him Adidas sports shoes. Thereupon a neighbour from our circle of helpers talked to him very intensively. He then developed into a son that no mother could wish for better.[14]

[12] Report on radio and television news on 16 January 2018.

[13] These experiences are reported here not only as examples of concrete problems in refugee assistance, but also to prevent the suspicion that the facts in this section are presented for Islamophobic/xenophobic/right-wing extremist motives.

[14] After graduating from the German school in Sarajevo, he was at the beginning of a promising career at a German financial institution in Bosnia. "Mom, now you don't have to work anymore," he told his mother. But on Mother's Day 2012, during a nighttime thunderstorm in the mountains of Tuszla, he was killed in an accident while driving home from his honeymoon.

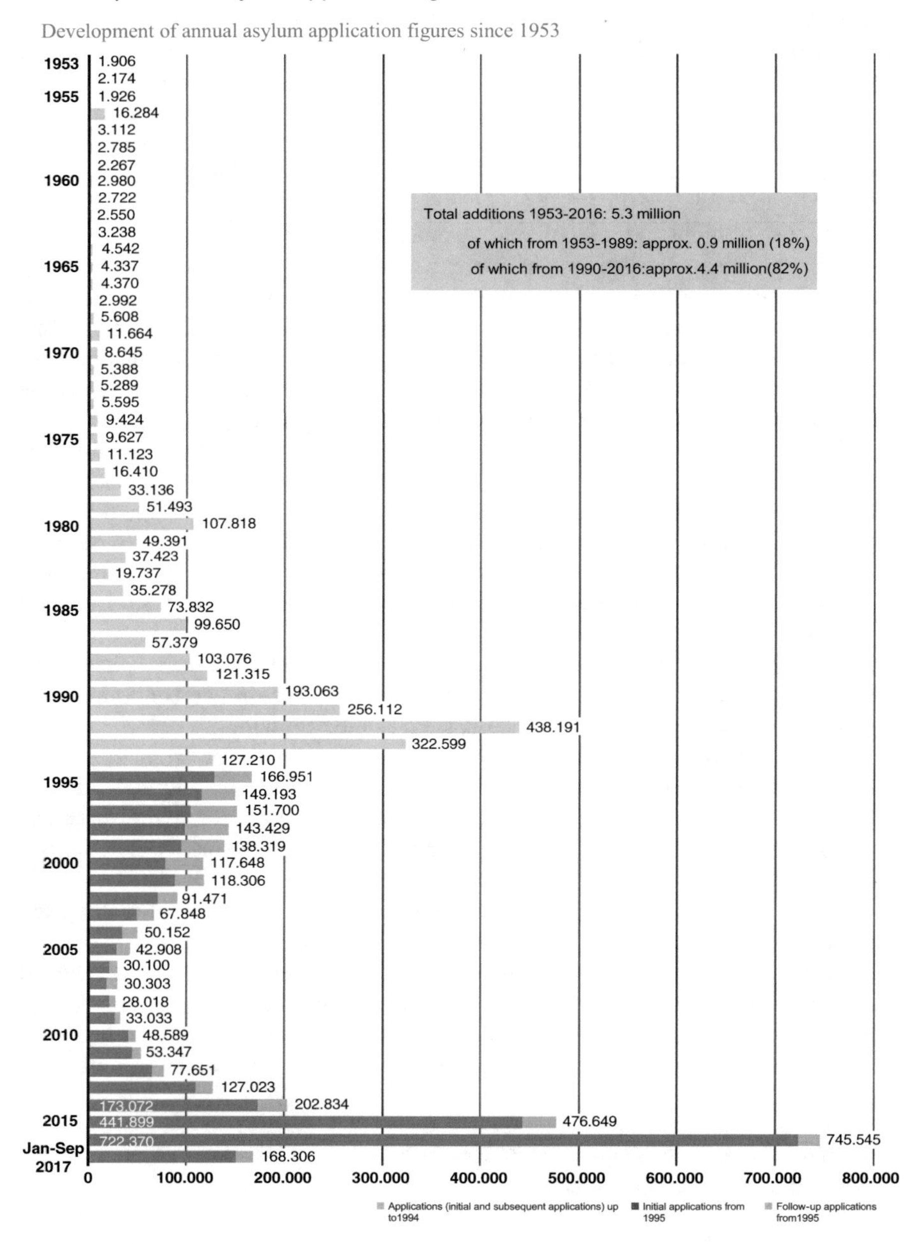

Fig. 5.2 Development over time of the annual number of asylum applications in the Federal Republic of Germany (FRG) since 1953 [192]

Table 5.3 Number of asylum seekers in Germany by country of origin with the highest number of arrivals, from 2013 to September 2017

Countries of origin	2013	2014	2015	2016	2017
Afghanistan	7735	9115	31,382	127,012	13,348
Albania		7865	53,805	14,853	
Eritrea	3616	13,198	10,876	18,854	8162
Iraq	3958	5345	29,784	96,116	16,088
Iran	4442			26,426	6675
Kosovo		6908	33,427		
Macedonia	6208	5614	9083		
Nigeria				12,709	5729
Pakistan	4101		8199	14,484	
Russia	14,887			10,985	3925
Serbia	11,459	17,172	16,700		
Somalia	3786	5528			5285
Syria	11,851	39,332	158,657	266,250	36,832
Turkey					5410

No entry means that the country was not one of the ten countries with the highest number of arrivals in the year in question [192, 193]

The second Bosnian family of four was friends with the first. Asked about their very high heating bills, which the German taxpayer would have to pay, the father of the family replied: "Everything that Germany spends on refugees is reimbursed to your country by the United Nations, after all."

Even greater illusions about Germany's resources and capacity in terms of its ability to absorb refugees and migrants have apparently prevailed since 2014—at least that is what Fig. 5.2 and Table 5.3 suggest.

Most of the countries of origin of asylum seekers with strong access, as shown in Table 5.3, are Islamic. This fits in with the fact that first Federal President Christian Wulff and then also Federal Chancellor Angela Merkel declared: "Islam belongs to Germany".

According to the latest figures from the EU statistics authority Eurostat, Germany alone took in far more than half of all refugees in the EU in 2016. According to the data, 445,210 asylum seekers received a positive decision in this country. By comparison, 710,400 people across the EU were recognised as needing protection in 2016. In second place in the ranking of countries that approved asylum applications was Sweden, with 69,350 positive decisions. Italy came in third with 35,450 admissions. However, if these figures also take into account the population of the respective receiving countries, Sweden lands in first place with 7040 recognised protection seekers per million inhabitants, Germany in second place with 5420 per million inhabitants and Austria in third place with 3655 per million inhabitants. (Source: tagesschau.de, accessed 29.04.2017). For the first half of 2017, Eurostat counted 357,625 asylum decisions made in Germany, of which 182,525 were positive, while in the other 27 EU states a total of only 199,405 decisions were made (source: dpa report, 05.12.2017).

The migration dynamics since the turn of the millennium, shown graphically and in tabular form, are related to a series of events that dampen the optimism that an era of peace and freedom for people and markets had begun after the end of the Cold War.[15]

After the terrorist attacks on New York and Washington flown by al-Qaeda suicide bombers on 11 September 2001, the US attacked Afghanistan, whose Taliban government had given sanctuary to al-Qaeda chief Osama Bin Laden and his followers. After conquering Afghanistan and installing pro-Western (and highly corrupt) governments that also enjoy the military and economic support of Germany, the US did not focus on bringing lasting peace to the country. Rather, in 2002, the US administration led by President George W. Bush sought authorization from the UN Security Council for a war of aggression against Iraq. The reason was that the dictator Saddam Hussein possessed weapons of mass destruction. Because of justified doubts about the existence of weapons of mass destruction, the authorization was not granted. After the bombardment of Baghdad in March 2003, US and British ground troops crossed the border into Iraq. France and Germany refused to join Bush's "coalition of the willing." On 1 May 2003, President Bush declared the war victoriously over. Weapons of mass destruction were not found. Earlier, German opposition leader Angela Merkel had flown to Washington and assured Bush that as chancellor she would have led Germany into the coalition of the willing.[16]

Then the Arab Spring threw the Muslim countries bordering the Mediterranean into chaos. Things got worst in Syria. Bush's coalition of the willing had not bothered with the weapons caches of the defeated Iraqi army. The Sunni supporters of Saddam Hussein, especially those dismissed from the army by the new Shiite government, helped themselves to them. Since then, attacks have continued to rock Iraq. In addition, a part of the Sunni minority joined the so-called Islamic State, which quickly gained strength in the second attempt. The latter murdered with other jihadists in the Syrian civil war. Some of the people uprooted by the war pushed across the Mediterranean, the Greek islands and the Balkan route, preferentially towards Germany. Iraqis, Iranians, Afghans and Pakistanis joined them.

In the summer of 2014, German President Joachim Gauck announced over radio and television that Germany was a rich country,[17] where refugees were welcome. After that, the number of Mediterranean crossings (and tragedies) by Africans also

---

[15]The surge of migration from Kosovo, which was less related to these events, was driven by a desolate economy, massive corruption and rumours of great prospects in Germany spread by criminal actors.

[16]There was a crude comment about this on a Düsseldorf carnival float.

[17]According to estimates by the International Monetary Fund in October 2016, Germany ranks twentieth in both international rankings of gross domestic product per capita—both in US dollars_{2015} and in international purchasing power parity. Ahead of Germany in the US dollar_{2015} ranking are Luxembourg, Switzerland, Norway, Macau, Qatar, Ireland, USA, Singapore, Denmark, Australia, Iceland, Sweden, San Marino, Netherlands, UK, Austria, Canada, Finland and Hong Kong. Of the total net wealth of German private households, 60% is accounted for by the richest 10% and 2.5% by the bottom 50% of households [194].

rose steeply. Refugees and migrants paid and still pay their smugglers many thousands of euros. The Somali boss of Africa's largest trafficking organization, who equips boat people with mobile phones to call European ships for help by distress call after leaving African territorial waters, became immensely rich, and the Roman mafia switched from drug trafficking and prostitution to the care and onward transportation of migrants and asylum seekers. On 15 May 2017, the anti-mafia authority in Catanzaro announced that the Calabrian mafia had made profits from managing a refugee camp in the province of Crotone—there is talk of EUR 32 million over 10 years. Organized crime has found a new, lucrative line of business.

Smuggling gangs now often use huge inflatable boats that are specially manufactured in China and delivered via the free port of Malta to the Libyan port city of Misurata, for example.

> The major Chinese trading platform Alibaba.com acts as a kind of Amazon for human traffickers. There, for example, the company Weihai Dafang from Shandong advertised various versions of a type with the trade name 'inflatable boat for refugees' – up to nine metres long, with space for '50 to 60 people'...on special offer, delivery time 30 days. [195]

When accessing the Alibaba website of the company Weihai Dafang 1 week after this press release from June 2017, the special offer "for refugees" had expired. Instead, the inflatable boats were offered for 50–60 people with a minimum order of 50 boats.

Even a regional newspaper like the Würzburg Main-Post, which strongly supports the Chancellor's refugee policy, lets dpa correspondent Petra Kaminski report on "The billion-dollar business of smugglers" [196]:

> UN migration expert Frank Laczko estimates that smuggling networks worldwide currently turn over ten billion euros a year ... The European police agency Europol estimates that with around a million people arriving in Europe in 2015, each paid an average of 3000 to 6000 euros to escape. Security circles suspect that 70 to 80 percent of the smugglers' turnover remains as profit. Payment is largely in cash. Money laundering and couriers who move large sums across borders are part of the system. According to Europol the money is invested in legal businesses such as car dealerships, greengrocers, real estate and transport companies. ... The arrivals remain silent when the authorities question them about the smugglers. Even if they have been mistreated on their journey. Many want to bring to Germany also friends and family – often with the help of the same traffickers. This is another reason why they rarely reveal the smugglers. The smuggling business is booming because of the demand for it', says Austrian expert Michael Spindelegger. ... Nine out of ten refugees use criminal help on their way to Europe. The demand arises for many reasons: Because war, violence and oppression ... are raging fiercely. Or because young men and women in Asia, Africa and the Caribbean are shown on large and small screens via the Internet what life chances they would have elsewhere. ... A complete set of forged or illegally procured documents for the journey to Europe, including a new driver's licence and birth certificate, can cost 5,000 to 10,000 euros. ... Wealthy people buy documents for 20,000 US dollars with a matching CV plus passage by yacht from Turkey to Italy. ... In Africa, security experts observe that smugglers are now advertising how to flee ... 'The smugglers are now looking for their own customers,' says UN expert Laczko. They go there and advertise their services on Facebook and other social media'.

Thanks to semiconductor advances in micro- and nanostructuring of transistors, tablet computers and smartphones, combined with the Internet, nowadays disseminate information in words and pictures about the good life of people in the highly industrialized islands of prosperity on our planet. They also enable money transfers from migrants who have made it to the industrialized countries to relatives in sub-Saharan Africa, the Horn of Africa, and Islamic Asia.

When someone in a Malian village builds a new house with money that his son has transferred home from Europe via smartphone, the neighbours and their relatives pool their scarce resources to also send one of their young men on the often deadly journey through the Sahara and across the Mediterranean to Europe by means of lavishly paid traffickers. Those who are not thrown out of their jeeps in the Sahara or taken hostage for ransom in Libya, and those who do not subsequently sink into the sea in an overcrowded boat that has sprung a leak or capsized by the passengers themselves, but are rescued by private or state rescue ships, make it onto TV in Lampedusa. It shows one announcing his arrival via smartphone. Then the next young man, designated by the clan chief, sets off on his journey.

Having arrived in Germany in October 2015, a minor told his volunteer helper, who had become a member of the Bundestag in 2017, that his relatives back home were now expecting real money from him and did not believe that the German state would only give him 300 euros a month to live on. That is why he has transferred 250 EUR to Africa and is trying to get through the month with the remaining 50 EUR.

According to the International Organization for Migration (IOM), the numbers of boat people arriving in Greece (GR), Italy (I), and Spain (E) were [197]

In 2016: 174,000 (GR), 181,000 (I), 8000 (E);
In 2017: 30,000 (GR), 119,000 (I), 22,000 (E).

Following the closure of the Balkan route by Slovenia, Croatia, Serbia, Macedonia and Hungary, and because of EU cooperation with Libya, migrant smugglers are increasingly diverting the flow of migrants towards Spain.

Until 2015, the German government had remained under cover behind the Dublin Agreement. But then, in September 2015, migrants stranded in Budapest's main train station set off on foot towards the Austrian border, with Germany as their final destination. Its citizens learned through television, radio and the press how, in response, Chancellor Angela Merkel announced Germany's open borders, especially for Syrian refugees, through all the world's communication channels, and that prices for fake Syrian passports then skyrocketed. The interior minister, who objected to the loss of control, was stripped of responsibility for border crossings. This responsibility was handed over to the chancellery minister.

German citizens reacted differently to these events. Some welcomed the refugees at the train stations and volunteered to help them in the many emergency shelters that were set up as quickly as possible.[18] Others turned to a new party on the far right of the party spectrum, which before September 2015 had been well on its way to disintegrating itself in wing wars. It then became the third strongest party in the 2017 federal election.[19] And many voters of the former major parties no longer know who they can trust to solve our country's problems.

With their involuntary delivery of advertising arguments for the gangs of traffickers, the Federal President and the Federal Chancellor had reacted to the welcome culture, which, oriented on the example of Sweden, had found and still has many supporters in Germany. The images and news about this German welcome culture and its state representatives, broadcast worldwide via smartphones and the social networks, created the myth of Germany as the promised land for all who dream of a better life, despite arson attacks on refugee shelters.

However, the onslaught of asylum seekers shown in Fig. 5.2 is now overwhelming the German asylum system and its employees.[20] The request of the prime minister of a federal state that the "gentlemen of the Federal Office for Migration and Refugees (BAMF) please work overtime even on weekends" and the hasty hiring of inadequately trained decision-makers could not prevent jihadists and even a right-wing extremist federal army officer from sneaking into the asylum and social system. Flaming indignation about this expressed on all sides does not help. Good administration is one of Germany's strengths. Damaging it by overtaxing it harms the country.

The administrative courts are also overburdened by challenges to the decisions of the BAMF, which, after hiring thousands of additional employees, has been processing asylum applications that have been pending for some time. Now, massive legal action is being taken against decisions that did not grant the status sought by the asylum seekers in question. While in June 2016 there were still around 68,000 asylum cases pending before German administrative courts, as of June 2017 the courts had already recorded around 320,000 cases.[21]

Reports by serious media about the high costs of establishing the identity and age of asylum seekers who have destroyed their passports, bogus reasons for asylum such as a threatened death sentence in the home country because of serious crimes allegedly but not actually committed there, Bogus conversions to Christianity and the groundless acknowledgement of paternity of children of asylum-seeking women by German men, to whom parts of the social benefits for the child and its mother are diverted by the latter, arouse more and more mistrust against the state, its elites and its institutions, which do not seem to be up to the migration problem.

---

[18] The Chancellor said: "It is part of our country's identity to achieve great things" [198]. EKD Council President Heinrich Bedford-Strohm echoed this: "2015 will go down in the history of our country as the year in which Germany surpassed itself" [199].

[19] The Chancellor's promise at the beginning of 2017 that what happened in the autumn of 2015 cannot, will not and must not be repeated did not help against this.

[20] The same applies to Sweden, which has introduced border controls.

[21] Main-Post Würzburg, 6 November 2017, p. 1: "We do 80% asylum cases"—Würzburg Administrative Court President on the burden on his authority.

With their exemplary voluntary efforts, the welcome culture creators have prevented chaos in the care and temporary accommodation of migrants at the height of the refugee crisis so far. But the integration of those who are allowed to stay here, which is invoked by everyone, can only succeed by providing housing in the middle of normal residential areas, and not in ghetto-like satellite settlements. This is where things have faltered in the past, and where they are faltering today. In white, urban middle-class neighborhoods on the East Coast of the United States, people fear a loss of property value if a black family moves near them. Liberal citizens of German neighborhoods harbor similar fears when homes for approved refugees are to be built near them. People are who they are. Trite public appeals by clergymen and politicians do not change this. Rather they may drive voters to the right-wing populists who are agitating against migrants in many EU countries and provide the smugglers with more advertising arguments.

EU members that used to belong to the Warsaw Pact and whose pensioners sometimes have lower incomes than recognised asylum seekers in Germany are securing their borders against the further influx of migrants. They also oppose redistribution quotas set by the EU Commission to reduce the imbalance in the distribution of asylum seekers within the EU. They are loudly rebuked for this, especially from Germany. After the UK's Brexit vote, in which immigration also played a role, the European Union is already sailing in troubled waters. German exhortations and simultaneous refusal to allow migrants taken on board from the Mediterranean by German-flagged ships to be taken to ports other than Italian ones are likely to put the EU at even greater risk. The anti-EU populists of the "Cinque Stelle" movement and the "Lega Nord" are profiting from Italy's migrant overload.

Needless to say, the official figures for asylum applications are an inadequate reflection of the influx of migrants into Germany and Europe.

Between 2014 and 2017, around 1.5 million refugees and migrants arrived in Germany. Examination and processing of their asylum applications, administrative court proceedings in the case of appeals against decisions by the BAMF, conflict mediation in their accommodation, integration and language courses, medical care, etc. are personnel-intensive. Personnel is expensive in Germany. By 2022, the federal budget has earmarked almost EUR 21 billion for social transfers to refugees, EUR 13 billion for integration services such as language courses, and EUR 5.2 billion for reception, registration and accommodation, reports the press [201]. But Germany is considered rich, and it can be seen as a balancing historical justice that our country, having never been presented with an official reparations bill from all the warring nations for the Second World War it started and lost, is now bearing the main financial burden of migration to Europe. More seriously, most immigrants have not received the education and training in their countries of origin that is required for gainful employment outside the low-wage sector in a highly industrialised country. Unless the language barriers that prevent immigrants from being trained as the skilled workers, which Germany urgently needs, are removed as quickly as possible, the production factor of labour is at risk of a long-term decline in quality. This is likely to become particularly problematic if this factor continues to be used to finance the common tasks of the state and the social security systems. Chapter 6 addresses this issue.

In Germany, young men make up around 80% of migrants. Single, without a partner or satisfying employment, young men of all races and religions are a problem. Germany became shockingly aware of this on New Year's Eve in Cologne in 2015. Migration incentives that raise unfulfillable expectations among billions of people via modern means of communication must be replaced by fighting the causes of flight.

A first measure on the part of the industrialized countries would be to reduce their tariffs against the import of products from the job-creating processing of domestic raw materials in the developing countries themselves. Their competition with products from the industrialised countries, e.g. European coffee roasters and chocolate manufacturers, would have to be accepted. And the fish piracy of non-African fishing factory ships off the African coasts should be stopped with European military aid.

But combating the reasons for flight caused by corrupt elites requires tougher measures. Section 1.2.2 proposes restrictions on the movement of capital. The sanctions mechanism is the inspiration for this, and has already been applied internationally, e.g. against Iran and associates of the Russian president. The basic idea is outlined again here.

The industrially highly developed states, in close cooperation with civil society organizations of developing and emerging countries as well as proven international non-governmental organizations, adopt a treaty to introduce a *capital flight brake. This* treaty provides that all banks in the states that are members of the capital flight brake agreement are prohibited, under the threat of license withdrawal, from accepting funds from natural persons and legal entities from countries with a level of corruption identified by "Transparency International" and classified as unacceptable by the states of the agreement, and from investing them, in whatever form, outside the home country of the person concerned. Trade between the corrupt countries and the rest of the world would be conducted through United Nations transfer banks.

Are civilian constraints not enough to eliminate the causes of flight, because kleptocratic rulers continue to drive the people of their countries into insurgencies and civil wars, one may resort to the option of robust UN peacekeeping missions, including years of "looking after" the pacified countries, as was done for Germany and Japan after World War II.

5.2 Colombia

The Enlightenment—"Man's exit from his self-imposed immaturity" (Immanuel Kant)—had, with the proclamation of human rights by the Declaration of Independence of the USA in 1776 and the French Revolution of 1789, also ignited the striving for progress and freedom and the struggle for independence from Spain in Colombia and Venezuela. But unlike in Europe and North America, the striving for progress did not also encompass the technical-economic sphere. Industrialization lagged far behind the development of art and culture. "The Protestant ethic and the spirit of capitalism" (Max Weber) were absent in Latin American countries influenced by Spanish Catholicism. But howsoever the value decisions that kept Colombia stuck in agrarian feudalism were influenced by religion, what was undoubtedly important for them was the diversity of climates, the geological makeup, the fertile nature and the history of the country.

The historical development is lamented by *Violencia*, the cumbia of violence:[22]

"Hay que fusilar unas treinta familias – You'd have to shoot about 30 families," Javier M. said with a friendly smile in the penumbra of the entrance to the Departamento de Física in front of palm trees and bougainvilleas under Cali's afternoon sun. I had asked my always amiable colleague at a Tinto how the dire social situation of broad sections of the population in the rich, beautiful country of Colombia could be changed most quickly.

His answer reminded me of our preparatory course by the Arbeitsgemeinschaft für Entwicklungshilfe in the autumn of 1969: "Colombia, with its highly educated upper and middle classes, was – and in some ways still is – the university of Latin America. But one can also observe that in a restaurant two elegantly dressed gentlemen discuss with each other in a highly cultivated manner until one pulls out a pistol and shoots the other," the country expert for Colombia had said, explaining to us that violence has been endemic in Colombia since the struggle for independence against Spanish colonial rule, which began in 1810 and was successfully concluded in 1819 with the victory of Boyaca.

Javier himself could never have harmed a human being. He was close to the Bahai faith and worked in his spare time in educational projects for the impoverished rural population. His answer was purely theoretical, because speed had been asked for. Younger, impatient people, on the other hand, thought the path of violence was viable and had joined revolutionary insurgency movements of various Marxist imprints. And all Colombians were then, in the early 1970s, and are now, only too painfully aware of the great tensions in their country. The motto of the Colombian national flag designates its poles: "Libertad y Orden – Freedom and Order".

Between these poles, tensions build up in all human societies, where the wealth gap between a few rich and many poor becomes too great. In managing these tensions, Colombia today stands at a bifurcation. The rest of the world is about to face a similar bifurcation.

Colombia and Germany have many ties. German fighter pilots of World War I founded Latin America's first civilian airline in Colombia after 1918. Brewers brought good beer to the country, and large companies set up branches. There are large German-speaking communities in Colombia's metropolitan areas. In German schools (*Colegios Alemanes*), supported by German tax money, German and Colombian teachers lead their students to the Abitur. Yet most Central Europeans

---

[22]The Spanish text of "Violencia" and its German translation cannot be reproduced here for technical reasons.

do not know much more about Colombia than that it is from this country that Western industrialized countries' demand for cocaine is satisfied, that it has problems because of this, and that for trying to solve these problems its President Santos, who left office in August 2018, was awarded the Nobel Peace Prize. Given Colombia's growing international importance, which Venezuela's disintegration reinforces, a little more in-depth information about the country and its people below will help us understand how and why creativity has worked and continues to work, also quite negatively, in South America's third-largest economy.

5.2.1 Official Information

The German Foreign Office [202] reports on Colombia in spring 2018:

Country name: Republic of Colombia—República de Colombia.
Climate: wide range of climatically different zones depending on altitude; predominantly tropical or subtropical; the capital Bogotá is in the temperate climate zone. Hardly any seasonal temperature fluctuations.
Location: In the northwest of South America. 1600 km of coastline on the Caribbean Sea and 1300 km on the Pacific Ocean. Neighbouring countries: Panama, Venezuela, Brazil, Peru and Ecuador.
Size: 1.138 million sq. km.
Capital: Bogotá, population 8.4 million.
Population: 49.4 million. Growth rate 1.2% per year. Composition according to 2015 census: 3.43% indigenous, 10.6% Afro-Colombian, 49% mestizo, 27% white.
Language: Spanish; 65 indigenous languages; English in San Andrés and Providencia.
Religion: Catholic (about 80%); increasingly Evangelical (about 20%)
National holiday: 20 July (Independence Day)
Independence date: 20 July 1810
Form of government: presidential democracy; Congress with two chambers: Senate and House of Representatives
Head of state and head of government: Juan Manuel Santos Calderón, since 7 August 2010, re-elected 15 June 2014, sworn in for second term 7 August 2014. Term of office 4 years. Next election May/June 2018.
Parliament: Constituent session of the parliament elected on 11 March 2018 on 20 July 2018. Future 108 seats in the Senate and 171 seats in the House of Representatives. The political party that emerged from the FARC is guaranteed five seats in both chambers for the coming legislative period under the peace agreement signed in 2016.
Governing party: Coalition of Partido de la U, Partido Conservador, Partido Liberal, Cambio Radical ... .
Gross domestic product: US$287.5 billion (2016)
GDP per capita: US$5897.5 (2016)

Country-Specific Safety Instructions
Since the Colombian government made peace with the guerrilla group FARC in December 2017, the latter has ceased its military activities. The demobilisation and disarmament phase has largely been successfully completed. However, FARC dissidents as well as other illegal organised crime groups continue to carry out attacks and gang warfare in the struggle for dominance in the power vacuum that has arisen. … The remaining guerrilla group ELN has also been conducting peace talks with the government since spring 2017. However, after a ceasefire agreement expired, targeted attacks on state facilities (police stations, power lines, oil pipeline) have been taking place since January 2018. The civilian population is not currently the target of such acts, but collateral damage is regularly accepted. Especially in the border regions of Colombia, as well as in rural, sparsely populated areas with weak infrastructure, state control is not guaranteed… .In the large cities (Bogotá, Medellin, Cali, Cartagena, Santa Marta) the security situation is comparable to that in other Latin American metropolises … .

5.2.2 Magnificent Nature

Colombia is one of the most beautiful countries on earth. It stretches[23] from the Leticia tip between Peru and Brazil at the headwaters of the Amazon at 4° south to the Caribbean sandy beaches of the Guajira Peninsula at 12.5° north, from the tropical rainforest of the Pacific coast at 77° west to the Rio Negro in the Colombia-Venezuela-Brazil triangle at 67° west.

Only sparsely populated is the eastern lowland of the *Llanos Orientales* covered by savannas and gallery forests and the adjoining tropical virgin forests, which occupies more than 50% of Colombia's national territory. The population is concentrated in the fan of the three mountain ranges Western, Central and Eastern Cordillera, into which the Andes split in the south of Colombia.

The 30–50 km wide mountain range of the Western Cordillera rises from the Pacific coast through tropical rainforest to more than 4000 m. It drops steeply down again into the valley near the city of Cali, which is about 30 km wide and at a height of 1000 m, through which the Rio Cauca flows northwards towards the Caribbean. *Alexander von Humboldt* described this fertile *Valle del Cauca* with its lush flora and fauna in perpetual summer as the most beautiful valley he was allowed to get to know on his travels. Its inhabitants call it "La sonrisa de Dios sobre la tierra – The smile of God above the earth".

After the narrow valley of the Rio Cauca, the Central Cordillera follows to the east with its glaciated, active volcanoes *Nevado del Huila, Nevado del Tolima* and *Nevado del Ruiz, which are* over 5000 m high. In it, at an altitude of 1500 m, is a basin containing Medellin, the city of perpetual spring and, with more than 2.7 million inhabitants, Colombia's second largest metropolis. The eastern flank of the

[23]Excluding the Caribbean islands of San Andrés and Providencia at the height of Nicaragua.

Central Cordillera ends in the valley of the mighty Rio Magdalena. Widening steadily, this valley descends from an altitude of 1600 m in the south to the north until it flows into the wide lowlands off the Caribbean coast. From this lowland, near the border to Venezuela and only 45 km away from the coast, the mountain range of the *Sierra Nevada de Santa Marta* rises mightily. Its two snow-covered peaks, *Pico Cristóbal Colon* and *Pico Simón Bolivar*, are both 5775 m high. The *Sierra Nevada de Santa Marta is* considered the highest coastal mountain range on earth.

Between the Magdalena Valley and the Llanos Orientales, the Eastern Cordillera stretches in places more than 200 km in width. There, on the eastern edge of a fertile plateau called *Sabana,* the capital Santafé de Bogotá is located at 2650 m above sea level in a magnificent mountain landscape. The highest elevation of the Eastern Cordillera is the *Sierra Nevada del Cocuy.* The western side of this mountain range grows out of the Andean highlands, its eastern side rises from the hot lowlands from 700 m through all climatic zones to 5330 m.

Thus, in the entire area of the Colombian Cordilleras, it is possible to pass through all climatic and vegetation zones of the earth on relatively short distances in a few hours—provided that rockfalls do not make the roads impassable. In it exist more than 20,000 species of plants, 1750 species of birds (20% of all bird species in the world), 1500 species of fish, 489 species of reptiles, 450 species of amphibians and 368 species of mammals [203]. In terms of biodiversity per unit area, Colombia ranks second in the world. Every time one travels through and flies over the country, one is always amazed by the beauty of Colombia.—It can well seduce one to devote oneself primarily to the beautiful things in life.[24]

5.2.3 Conservatives Versus Liberals

The Colombians tell their special form of the creation story with grim humour: 'When God had created the world and was looking at Colombia, he said: 'This land is too beatiful. It rivals heaven. It needs a dark side.' And he created the Colombian."

Simón Bolivar: The Liberator Who "Ploughed the Sea"

As great and versatile as Colombia's nature was the "Father of the Fatherland" *Simón Bolivar.* Born on 24 July 1783 in Caracas, Venezuela, he died on 17 December 1830 near Santa Marta, Colombia.

In his book *Der Marschall und die Gnade* (*The Marshal and the Grace*), published in 1954, Kasimir Edschmid recounts

> the unusual life of Simón Bolivar, who liberated South America from Spanish rule … .He evokes in his work the genius of the mysterious, vast continent which, at the beginning of the 19th century, was the ardent, feverish, bloody, torn and shaken scene of human passions,

[24]A Colombian physicist with a high international reputation, who had done his doctorate in Germany and research in the US, said of scientists in Colombia, "Lo peor es un tipo que se ha tropicalizado." [The worse is one who has adopted tropical behaviors].

ideals and errors. It sets an enduring monument to the life and deeds of its bold revolutionary. The campaigns that Simón Bolivar led are of Caesarian stature. But Bolivar, the revolutionary and republican, was a far-sighted politician of a different caliber

Spain, with unbridled and brutal terror, ruled most of South America in the early 19th century, countries twenty times the size of the mother country. Simón Bolivar came from the rich aristocracy of these colonial countries, he knew Europe, he had friends there, and a fantastic, enjoyable life lay ahead of him. But he did something unusual: he freed thousands of his Negro slaves and led an unparalleled war of liberation – in Venezuela, in Colombia, in Ecuador, in Bolivia and in Peru. Kasimir Edschmid tells with great urgency of the inner struggles of Bolivar, one of the most fascinating figures in history, who dedicated his life and work to democracy and the freedom of his people. The author places him within the ever-present problematic of the ancient questions of might and right, grace and being chosen, politics and conscience, anarchy and dictatorship. Bolivar was a writer, orator and general; he was Rousseau, Lord Byron and Napoleon all rolled into one. He was no enthusiast; he was a man of indomitable toughness, though of feeble health, a lover of women, and a man of pleasure in the midst of an existence in which paradise and hell were mingled. He had a grand conception of a South America which should unite almost all the states of that continent. He realized this dream – and saw it shattered [204].

35 years after Kasimir Edschmid, the Colombian Nobel Prize winner for literature *Gabriel Garcia Márquez* describes in his novel *El general en su laberinto.* [*The General in his Labyrinth*] the last voyage of the seriously ill 47-year-old Bolivar, accompanied by a few loyal followers, on the Magdalena River towards the Caribbean and possibly exile. In flashbacks he lets the glorious liberator appear once again and at the same time shows him in the labyrinth of his sufferings and lost dreams. For over 10 years, Simón Bolivar had experienced the heights of glory but also the downs of corrupting power and the attacks of his opponents as president of the state of Colombia he created. In 1830, in Santafé de Bogotá, he declared his resignation. "He who dedicates himself to a revolution ploughs the sea," he said bitterly at the end of his life [205].

Civil Wars

Freedom and self-determination did not bring peace to Colombia. Nine civil wars swept the country in the nineteenth century. The following sketch of the armed conflicts[25] between the supporters of the liberal and conservative parties shows how a republican-agrarian feudal society shaped by different values is torn apart by struggles that were also always about the soil, which converts solar energy into wealth and power through the photosynthesis of its plants.

1. Even during the liberation struggle against the Spanish, the centralists under Antonio Nariño and the federalists, since 1814 under the leadership of Simón Bolivar, fought for power. The centralists wanted a central government based in Bogotá, the federalists wanted to organize the country on a federal basis. Even

[25] The sketch of the civil wars is based on the essay: *Cronología de las guerras en Colombia—Guerras civiles de Colombia* [206].

though they eventually made common cause against the Spanish, Colombians refer to this first period of internal strife as “patria boba – foolish fatherland”. After the expulsion of the Spanish from Colombia and large parts of Venezuela, the Republic of Greater Colombia, consisting of Colombia and Venezuela, was proclaimed in 1819 and Simón Bolivar was elected its first president. The republic was joined by Panama in 1821 and Ecuador in 1822. In 1830 Venezuela and Ecuador seceded again. Colombia and Panama formed the Republic of New Granada (until Panama declared itself independent in 1903 under pressure from the United States).

2. The so-called “Guerra de los Supremos – War of the Supreme Commanders” of 1839–1841 was a failed rebellion of local caudillos in southwestern Colombia against the central government. Each of these caudillos considered himself commander-in-chief of his private army. The war, named after them, was fought for the purpose of redistributing economic and political power and was lost by the insurgents for lack of orderly command.
3. The Civil War of 1851 was instigated by conservative landowners in the south of the country. They fought the reforms of the liberal central government. The reforms involved freeing slaves, expelling Jesuits, abolishing the death penalty, freedom of the press, and jury justice. The government was victorious.
4. In 1854, General Melo staged a coup against President Obando, remained in power for 8 months, and was then defeated by an alliance of sections of the Liberal and Conservative parties after bloody fighting.
5. In the Civil War of 1860–1862, liberals rose up against conservative President Mariano Ospina Rodriguez. It is the only conflict in which the insurgents were victorious. They changed the constitution from centralist to federalist.
6. The Civil War of 1876–1877 was waged by conservatives against the liberal government to stop the anti-religious and anti-clerical, secular reorganization of education. The Catholic Church supported the rebellion. Guerrilla units formed by militiamen harassed the peasant population severely.
7. In the Civil War of 1884–1885, the radical wing of the Liberal Party rose up against the centralist reforms of President Rafael Núñez’s government. The moderate wings of the liberals and conservatives supported the president. After the victory of the government forces, a new constitution was promulgated in 1886. It created a central state with departmentos as administrative units. This constitution, which was reformed several times, was in force until 1991.
8. In 1895, the militant faction of the Liberals organized a failed uprising against the conservative president Miguel Antonio Caro.
9. In the “Guerra de los Mil Dias – War of 1000 Days” from 1899 to 1902, the Liberal Party attempted to overthrow the government. In this war, the most severe and consequential to date, poorly trained, anarchic liberal guerrillas fought against the well-organized government troops. Results of the civil war were the victory of the conservative party as well as a prolonged period of its hegemony, more than 100,000 deaths, and the economic disruption of the country. The latter made it easier for the United States to pursue the secession of Panama. This took place in November 1903.

In the first half of the twentieth century, Colombia experienced a period without armed political power struggles until 1945. (Perhaps Europe's self-destruction in the two world wars exerted a moderating effect). In 1921, the US paid US$25 million in compensation for the loss of Panama. This boosted the recovery of an economy built on coffee and banana plantations and the development of oil wells. The recovery suffered a severe setback due to the fall in the world market price for coffee as a result of the Great Depression from 1929 onwards. The conservatives, who had been in power until then, lost power to the liberals, whose president Alfonso López Pumarejo introduced social and agricultural reforms in 1934. It was not until 1946 that the conservatives returned to power with President Mariano Ospina Pérez. While in the capital liberals and conservatives were initially still on good terms with each other—political power in the two parties was in the hands of about 30 interconnected large families—conservative landowners exacerbated tensions in the countryside. The acts of violence in which these were unleashed were eventually also increasingly felt in the cities.

On April 9, 1948, at noon, Jorge Eliézer Gaitán, a 45-year-old social reformist Liberal presidential candidate, was shot dead by a 26-year-old assassin in downtown Bogotá. An angry crowd lynched the assassin and rapidly grew into a spontaneous popular uprising, the *Bogotazo*, which destroyed 142 buildings in central Bogotá. The conservative government responded with ferocity. The uprising spread to other cities and caused damage nationwide estimated at US$570 million [207]. It opened the bloodiest period in twentieth century Colombia, which dominates the memory of Colombians as *La Violencia.* The fighting between liberals and conservatives, conducted with extreme ferocity, claimed between 200,000 and 300,000 lives and resulted in more than two million internally displaced persons, out of a population of 11 million at the time.

In 1953, General Rojas Pinilla took power in a bloodless (!) coup d'état. He ended the first phase of the Violencia with dictatorial measures. A populist reformer similar to Argentina's Juan Peron, he initially won many sympathies. However, press censorship, suppression of all opposition, and economic difficulties then turned the middle and upper classes against him. A broad strike movement forced him to resign in 1957. He handed power to a military junta. In 1958, conservatives and liberals buried the hatchet. Replacing the military junta, they agreed to exercise power together as the *Frente Nacional* for 16 years. During this time, presidents alternated between the Liberal and Conservative parties, regardless of the outcome of elections. This ended 150 years of armed conflict between conservatives and liberals. But for the internal peace in the country, the unification came too late.

Marxist-influenced insurgency movements had formed out of the tormented, uprooted masses of the population. They were directed against the oligarchic elites of the political castes in the two traditional parties that had ruined the country. The basis of the economic and consequent political power of the elites was still the ownership of land. Due to the civil wars and a disregard for technical and scientific education, Colombia, like other countries in Latin America, had not succeeded in freeing itself from the shackles of feudalism and finding its way into industrial society.

In his article "Los Irresponsables – The Irresponsible Ones", published in 1972 in *El Tiempo,* Colombia's leading national daily [208], *Antonio Caballero* explains the disinterest of the Latin American upper classes in the fate of their peoples, as well as their inability to solve the social and economic problems of the countries they govern, through their selfish fixation on land ownership:

> El patriotismo de las clases altas latinoamericanas no es nunca amor a la nación, sino amor a la tierra; el pueblo es siempre, de los otros.
>
> [The patriotism of the upper classes of Latin America is never love of one's own nation, but merely the love of land; the people are always "the business of others."]

5.2.4 The Explosive Power of Population Growth Without Industrialisation

Uninterested in industrialization and the common good, Latin America's relatively few wealthy invested (and still invest) only a small portion of their vast fortunes[26] in the technical and economic development of their countries. They much preferred to invest their money in banks of North America and Europe and enjoy its interest earnings in a life of educated idleness surrounded by European and North American luxury goods. As late as 1972, Colombia was still paying licenses for technical processes whose patents had expired 50 years earlier. Only no one bothered to use the unprotected patents to develop an independent process. This dependence on foreign countries created too few industrial jobs for Colombia's (and other Latin American countries') population, which had been growing rapidly since the 1930s. Thus, most people were left with low-paid agricultural wage labor for the large landowners or rural flight to the cities, hoping to find odd jobs and services for the narrow middle and upper classes. The rural refugees found shelter on plots of land on the outskirts of the big cities, which they occupied in the evening and built overnight with makeshift huts made of cardboard boxes, sticks and plastic sheeting. If they were able to resist the police's attempts to evict them for long enough the next day, they had won a permanent right to stay. From the 1960s onwards, the "Barrios de Invasión" created in this way, with tens of thousands of inhabitants, allowed the large cities to grow far out into their surrounding countryside. The shabby shacks from the night of the invasion were replaced as quickly as possible by sturdier sheds and little houses made of slats, corrugated iron, and bricks. In front of them, people planted small gardens with plants whose bright flowers at first glance hide the misery of the dwellings, the muddy streets and the stinking, open sewers. In these barrios, the "normal" crime of theft, robbery and manslaughter is rampant.

The impoverishment of broad sections of the population, combined with Colombia's special climatic and geographical conditions and the demand from North America and Europe for cocaine, heroin and marijuana, drove the country

[26] See the number of so-called *High Net Worth Individuals—HNWI* and their wealth in international comparison, e.g. in [2, p. 230].

into a period of murders and kidnappings towards the end of the 1970s. These were committed by "left-wing" guerrillas, "right-wing" death squads and the nationally and globally well-connected, technically highly equipped criminal gangs of the drug cartels that cooperated with them. In terms of cruelty and the number of people murdered, the period of the civil wars was far surpassed. According to the Colombian police authorities, almost 32,000 people were murdered nationwide in 1997 alone, most of them in the three major cities of Bogotá, Cali and Medellin. Even in 1999, Medellin and Cali still exercised the greatest caution with their visitors in public. For quite some time, Colombia was considered the most dangerous country on earth.

The main actors in the fighting in the country, which has always been ruled by civilians except for about 6 years of military rule, were and are guerrilla units, the national armed forces, as well as drug cartels and paramilitaries.

The FARC, *Fuerzas Armadas Revolucionarias de Colombia—Revolutionary Armed Forces of Colombia*, were officially founded in 1966. They emerged from the union of small farmers and rural school teachers for the purpose of self-defence against the marauding gangs in the times of the *Violencia*. The FARC, who described themselves as Marxist, were the largest guerrilla organization in Latin America. With retreats in the jungles and mountainous regions, they initially fought to improve the economic situation of the rural population. To finance their activities, they later cooperated with the drug mafia and developed kidnapping into a lucrative business. The ransom income and profits from drug cultivation and trafficking allowed the fighters to be paid, which made the FARC attractive to young unemployed people. After the peace agreement with the government, the FARC's approximately 7000 male and female fighters must be integrated into civil society. This requires retraining them for civilian jobs and creating jobs with an income that is not significantly less than the previous pay of a fighter. This poses major challenges for Colombian society.—Perhaps a leap into the post-growth economy would be one, or perhaps even the answer: the high import tariffs on industrial consumer goods in the second half of the twentieth century had awakened impressive improvisational skills in the population, and especially in the Colombian middle class. Building on this, the promotion of craft skills within regional economies, as discussed in Sect. 4.8.7, could create livelihoods that would not require costly investment in high-tech production capacity.

The ELN, *Ejército de Liberación Nacional—National Liberation Army,* had been founded in 1964 by students as a Marxist-oriented guerrilla movement, was strengthened by guerrilleros from the *Violencia* period, and is one of the oldest guerrilla organizations still active in Latin America. It encouraged peasant revolts. Catholic priests such as *Camilo Torres* joined it. After suffering heavy military defeats in the early 1970s, the ELN reorganized in the 1980s, thanks in no small part to ransom payments of several million US$ by Germany's Mannesmann AG for the release of four employees kidnapped by the ELN during oil explorations on Colombia's Caribbean coast. Peace negotiations with the ELN, among others in the Würzburg monastery Himmelspforten, have so far remained without result. FARC fighters who reject the peace treaty with the government join the ELN.

Two smaller guerrilla groups formed after 1967, EPL and M19, no longer play a role.

The armed forces formed by the army, navy and air force consist of conscripts and professional officers; the latter are often trained in the US or by US military personnel. The armed forces bear the brunt of the fighting against insurgents and drug cartels. The military police also perform law and order duties in the interior, such as clearing and occupying universities whose students revolt, or securing airports.

In the 1980s and 1990s, the Cali and Medellin drug cartels, fighting the state and warring with each other, terrorized the country with bomb attacks. After the two cartels were dismantled, either through operations by the armed forces that ended with the shooting or capture of the bosses, or by threats of extradition of the clan chiefs to the United States if the cartels did not cease their "operations," drug production and trafficking passed into the hands of many small, hard-to-control criminal gangs that bought the support of the FARC. Large landowners, frequent victims of kidnappings by FARC and ELN, organized paramilitary units not only for their protection but also to enforce their economic interests. These not only fought the leftist guerrillas, but also murdered small farmers and trade unionists. They treated human rights activists and clergymen in the same way when these defended the interests of the little people against land theft and oppression.

Hernan Echavarría Olózaga (1911–2006) describes the disastrous consequences of the stagnation of industrialisation in feudalism from the point of view of social and economic policy. In liberal governments he had headed the ministries of "Obras Publicas – Public Works" (1943) and "Comunicaciones – Post and Telephone" (1958). He was one of the harshest critics of President Ernesto Samper Pizano (1994–1998). He had demanded his resignation because of donations from the drug mafia for the presidential election campaign. In the third chapter of his book *Miseria y Progreso (Misery and Progress)* [209] he relentlessly criticizes the wealthy Colombian upper class, to which he himself belonged.

At the outset, the author asks, "What happened in Colombia and other nations of Latin America during the nineteenth century, as Europe and North America industrialized and opened up new periods of economic, social, and cultural development?" and answers, "Frankly, not much, in any of those areas." He cites feudalism as the reason. His reckoning with it is summarized as follows.

The Industrial Revolution was an exclusively European development in the nineteenth century, but it quickly spread to North America. The peoples of Latin America had to wait until the twentieth century to participate in global technological progress. It was finally made possible for them in some areas through the development of transportation and communications.

During the colonial period, the feudal system of production prevailed throughout Colombia, as in many parts of the world. The land belonged to royal plenipotentiaries and was worked by peasants, the majority of whom were indigenous. Even after the end of Spanish rule, the production system remained feudal. Now it was the descendants of the royal plenipotentiaries who had the land worked by peasants. And to this day Colombia remains in feudalism. The few Spaniards who arrived as immigrants did not bring any significant manual skills with them because, among other things, in Spain these were assigned to "lowly mechanical" work that could not be expected of people of noble descent.

Influenced by the feudalism of the nineteenth and large parts of the twentieth century, the ruling class of Colombian society preferred to devote itself to literature and art, and this quite successfully, as well as scholastically influenced science. But entrepreneurial activities were of less interest, so that Colombia had not been able to participate as other parts of the world in the enormous technological upswing since the end of the Second World War. Nevertheless, modernization has taken place since 1945, largely under the influence of the United States, whose interest in Latin America had been rekindled. The reconstruction of devastated Europe was also exemplary. It showed the Latin American ruling class what economic progress was possible within a short time in industrial societies.

The economic policies of Colombia, like those of Latin America, fluctuated in the second half of the twentieth century between more or less neoliberal market economies and attempts to go their own way, trying out inflationist, planned economy and autarkist theories. The results were miserable, as they always are when ideology and folly arbitrarily change the framework of the economy or prevent adjustments to the technical constraints of industrial production. For this reason, and because of the feudal mentality of the ruling class, which has still not been overcome, Hernan Echavarria Olozaga fears that Colombians will have to wait some time for the achievements of the Industrial Revolution and the social advances and personal freedoms that go with them.

But the spiritual and intellectual conditions for a change for the better are becoming visible in the fields of faith and natural science.

5.2.5 Suffering and Change

On 16 March 2002, the Archbishop of Cali, *Isaías Duarte Cancino*, was shot dead by teenage contract killers after the wedding of 100 couples in front of the church of the poor district Aguablanca in Cali. Like Oscar Romero, Archbishop of San Salvador, murdered on 24 March 1980, while celebrating Mass, he represents change in the Catholic Church in Latin America and its option for the poor.

Born on 15 February 1939 in San Gil, Santander district, the youngest of seven children, he studied theology in Pamplona and Rome from 1956 to 1964. He was ordained priest in 1963. Returning to Colombia, he worked in Bucaramanga. From 1985 to 1988 he was Auxiliary Bishop there, and from 1988 to 1995 Bishop of Apartadó in the province of Urabá. In that region guerrillas and paramilitaries had at that time started those cruel fights which still continue today and in which the campesinos again and again get between the fronts and are killed. There Isaias Duarte's conception of the exercise of his priesthood changed forever.

After he had been installed in 1995 as archbishop of the archdiocese Cali, he stood up in word and deed for peace and justice and tried to remove the breeding ground of violence by social works in the poor quarters.

After his murder, the perpetrators were quickly caught and died soon after. The FARC, the drug mafia and government agencies were and are suspected of being behind the murder. In the weeks before his death, Isaias Duarte had denounced political corruption and its connection to the drug mafia with all clarity and severity and preached against the violence of guerrillas and paramilitaries.

Close associates of Isaias Duarte feared for their lives and fled abroad. In view of the indignation at home and abroad the state offered all Colombian bishops personal protection. This was refused by the Colombian Bishops' Conference on the grounds that the bishops wanted to live with the same risks as all other believers. This is all the more remarkable because in the early 1970s the Colombian clergy was considered the most reactionary in Latin America.

At that time only a few foreign clergy were active in the poor quarters of Cali with hundreds of thousands of inhabitants, which had been created by land occupation. A bishop did not allow himself to be seen there. "He was afraid that the whitewall tires of his Volkswagen car would get dirty from the mud of our streets," the people told German partners of four Spanish religious who were building a church with an educational and social center in the poor district of "El Rodeo."

At that time, the social elite was indifferent to the impoverished peasants streaming into the slums of the big cities—apart from buying votes in elections. Protection from the growing crime was sought from private security services. But these could not prevent murder and kidnapping by the drug mafia, guerrilleros and paramilitaries from reaching even the wealthy middle and upper classes. That is when a turnaround began: the elite's turn to the people.

Now, in the slums still ravaged by fighting from armed gangs, many local priests and religious sisters, along with members of the higher and highest levels of society, are developing basic services of food, sanitation, medicine, education, legal counseling and church services. In doing so, the laity use their professional experience as entrepreneurs, managers and administrators, teachers and educators, medical professionals and psychologists to expertly minimize inefficiency and abuse. As a result, programs such as the "Banco de Alimentos" (equivalent to the German "Tafeln" for food distribution to the needy) are self-financing during operation. At the same time, by presenting the projects in secondary schools and raising awareness among children and young people, attempts are made to win their parents over to voluntary work or financial support for new programmes.

The change in the Colombian elite is also noticeable in technical and scientific progress, which since the mid-1960s has been preparing the ground on which Colombia's industrialization, adapted to the country's needs, can develop, provided that policy-makers set the right course. The principles of the post-growth economy mentioned in Sect. 4.8.7 in connection with the demobilisation of the FARC could help to set this course, avoiding the ecological mistakes made by the highly industrialised countries and generating income for the rapidly growing young population. If Colombian consumers were to become prosumers in accordance with Sect. 4.8.4, they would now have a wealth of imported goods from Chinese production at their disposal.

Technical and scientific progress began with the sending of capable, dynamic young engineers and physicists to North America and Europe for advanced studies and the invitation of lecturers in these subjects from these same regions to Colombian universities. Initially, Colombian students had to overcome mental barriers erected by a traditional education system focused on rote learning. But once that was done, the intelligent, inquisitive, highly motivated students in the physics master's programs at the Universidad Nacional in Bogotà and the Universidad del Valle in Cali performed as well as the best physics diploma students in Germany. After graduation, they went abroad to earn their doctorates, returned, and in hard work at universities, in scientific organizations, and in laboratories in the private sector, they raised technical-scientific teaching and research to a level at which publications in the leading international scientific journals thrive. In doing so, they do not remain in the academic ivory tower, but face the problems of their country. Whether they will be able to use their skills to sustainably improve the living conditions of the poor population will depend on whether the thin but immensely rich Colombian upper class will bring its financial resources back to Colombia from abroad and invest them in the country's industrial development.

Colombia appears as an image of our world growing together. In this one country, as on our planet, we find the diversity of races, classes, cultures, ecosystems and climates; likewise the diversity of human behaviour: from courageous commitment to the liberation of fellow human beings from poverty and violence to murder for economic gain and political power. In this land, Homo sapiens unfolded his potential for the noblest as well as the worst actions in the absence of a functioning state monopoly of violence, committed to human rights, which would be able to develop and sufficiently restrict violence for the enforcement of economic and political interests. Now, however, the pressure of suffering is transforming Colombia's elite: social solidarity, sober work, scientific-technical creativity and the fight against corruption are gaining a status they did not have before.

Will Colombia, and likewise the world community of states, succeed in limiting ruthlessly selfish human aspirations through law and order and powerful, corruption-free institutions sworn to this end? And how can the industrialization of the developing and newly industrialized countries, from which the swelling flood of migrants is pushing into the industrialized countries, succeed without destroying the natural foundations of life? Will human creativity find the guard rails of future economic development within which the tension between freedom and order can be sustained and the gap between rich and poor narrowed?

6 What Will We Choose?

Between the discovery of America and the end of the Second World War, Europe had an ever-increasing influence on the destinies of people on all continents. First, firearms and sailing ships put the chemical energy of gunpowder and the kinetic energy of wind ever more efficiently at the service of European military and economic expansion. European feudal lords exploited the soil and people of the colonies. The colonies of the American double continent broke away from their mother countries in the eighteenth and nineteenth centuries. But the epochal change that, since the Industrial Revolution with heat engines, had harnessed natural energy sources to provide the energy services that made economic and population growth possible, was long reserved outside Europe for North America and Japan. Yet Japan had its first contacts with Europeans and their firearms in the mid-sixteenth century. After a short period of cultural, economic and military-technological exchange, it again completely closed itself off to European influence and, in order to preserve the military samurai tradition, also gradually discontinued its impressive production of excellent firearms at the beginning of the seventeenth century. It was not until 1853 that Commodore Perry's "visit" with a cannon-wielding U.S. fleet convinced Japan of the advantages of opening the country to European-North American influences. Firearms production was resumed and industrial production started up so successfully that the resulting rivalry with the USA only ended after the latter dropped atomic bombs on Hiroshima and Nagasaki on 6 and 9 August 1945.

In the twenty-first century, there are no more colonies in the political sense. But the rest of the world cannot and will not do without the industrial production developed by Europe and the economic and political strength associated with it. The fact that the whole world is also interested in the arms industry is one of the problems that must be expected to worsen due to the consequences of the anthropogenic greenhouse effect.

In their industrial development, Latin America and Africa seem to be catching up with the early industrialized countries more slowly than the countries of the Eurasian landmass. In the case of Colombia, and other parts of Latin America, feudal

R. Kümmel et al., *Energy, entropy, creativity*,
https://doi.org/10.1007/978-3-662-65778-2_6

structures inherited from agrarian societies, combined with capital flight, are probably responsible for this. The conditions described in Sect. 5.2 suggest as much. It seems rather unlikely that geographical and climatic barriers to the diffusion of innovation, as in the days of advanced agrarian civilizations [200], still play a role in the age of air travel and the Internet.

Having subjugated the world to their techno-economic influence, the highly industrialized market economies, first and foremost Europe and North America, have contributed massively to the increase in atmospheric greenhouse gas concentrations by raising the standard of living of their populations. It is now appropriate, indeed imperative, that they take the lead in overcoming the economic and ecological problems in international cooperation. The less often national, inefficient special paths are taken, the better.

In the following, we consider three scenarios of economic development. They take into account the constraints imposed on industrial production by the Second Law of Thermodynamics. The likelihood that one of these scenarios will be realized is left to the reader's judgment.

Under the motto "Sovereign is not he who has much, but needs little" [210], the first scenario assumes thorough changes in human behaviour. They are the prerequisite for a low-conflict coexistence of presumably 10 billion people in a stationary world society in which economic growth no longer helps to avoid distributional struggles. Niko Paech describes this in Chap. 4.

The other two options concern financial and space developments.

Since the entropy production associated with energy use limits the growth of value creation within the biosphere, the question arises whether and how fiscal instruments can be used to redistribute the value creation owed to energy between the state and the private sector in such a way that economic dynamics and thermodynamic constraints do not lead to intolerable social tensions. If this succeeds, resources for overcoming the limits to growth may also be found [211].

6.1 Rerouting the Economy via Energy Taxation

Even the relatively simple examples of energy, emission and cost optimisation from Sect. 2.3.3 show that in an economy that aims to minimise costs, higher energy prices are more likely to lead to energy savings and emission reductions than low ones. Moreover, in industrial production, according to the econometric analyses of Chap. 3, the (still) cheap energy that activates capital is much more powerful in terms of production than expensive human labour. Therefore, as digitisation progresses, people employed in routine work will be pushed out of the value creation process into unemployment by growing automation, if economic growth slows down and no new jobs are created in newly emerging sectors of the economy.

For November 2017, the Federal Employment Office calculated a German unemployment rate of 5.3%. The term unemployment, which is defined in the social laws, has been increasingly narrowed with each amendment of the law. If one adds to the unemployment rate the rate of the so-called underemployed, which includes older, unemployed Hartz IV-welfare recipients, the total rate for Germany is 7.4%. Nevertheless, since the first global economic crisis of the twenty-first century was overcome, industry and the skilled trades have been complaining about the worsening shortage of skilled workers. The handling of apparatus and tools, which are becoming more and more complex with the increasing use of energy, requires an ever higher level of training, to which many older people who have become unemployed due to automation are no longer able to rise. Young migrants are hindered by language and other educational deficits.[1]

It is uncertain how long energy will remain as cheap as it has been since, say, 1950, despite all the fluctuations in the price of oil. *The 1950 Syndrome*, the publication of an interdisciplinary research project at the University of Berne, makes the following observation about the price of energy:

> The post-war boom is currently being discovered as an era of fundamental changes that are shaping our society today. Until around 1950, Europe moved along a relatively environmentally compatible development path. It was only in the following decades that energy consumption, gross national product, the amount of land required for settlements, the volume of waste and the pollution of air, water and soil experienced the growth spurt that is decisive for today's undesirable development. The thesis of the '1950 syndrome' ... considers the long-term decline in relative energy prices as one of the main driving forces of the development so far. [212]

In view of the intended phase-out of coal, the "peak oil" prospects for conventional oil and gas reserves [213], the cost and environmental problems of non-conventional fossil energy resources, and the sharply rising debt of many oil and gas producers,[2] the "essential drivers of development to date" may soon come to a standstill.[3]

It is also uncertain how long Germany will continue to fare better than many of its European neighbours in terms of unemployment. After all, China will soon be competing with Germany on the world market to satisfy the demand of the industrialising developing and emerging countries for means of transport, machine tools and industrial plant.

---

[1] A strategy of remedying Germany's shortage of skilled workers through immigration of skilled workers from developing countries would deprive these countries of a prerequisite for improving their living conditions. Only if Germany approves applications from developing countries to help them establish and finance educational institutions in the fields of agriculture, handicrafts, technology, and medicine, and offers language training and a job in Germany to some of those trained in them, would Germany's benefit not be bought at the expense of the developing countries.

[2] The latest research on global oil supply shows that conventional oil production peaked about a decade ago. The amount of energy that must be expended to provide one unit of oil, called *Energy Returned On Invested (Energy), (EROI) is* growing. The number of newly discovered oil fields is falling dramatically, while capital investment in exploration and production has increased tremendously. Fracking and other methods of tapping unconventional sources of oil and gas will not change this much. The export capacities of many oil-producing countries are shrinking rapidly because of growing internal problems [214].

[3] As of November 2017, the US national debt, as defined as that of the federal government in Washington alone, was 106% of GDP.

High social and environmental standards can be a competitive disadvantage. However, they can also become a competitive advantage if they are successfully introduced and further developed in one's own economic area. One instrument that has been discussed since the 1990s is the **shifting of the tax and levy burden from labour to energy**. In this context, "energy" always refers to the *exergy content of* a quantity of energy.

6.1.1 Taxation According to Performance

In its monthly report of August 1997, the Deutsche Bundesbank demanded: "The focus of a tax reform would have to be the fundamental reform of income taxation. ... An important component of such a major reform is a certain shift of the tax burden from income to consumption." And in the same context, the Federal Ministry of Finance recalled the "proven basic principles of income tax law, in particular ... (the) principle of taxation according to economic capacity as a ... fundamental principle of fair taxation" [215].

The Constitution of the Federal Republic of Germany requires taxation according to economic performance. In view of the imbalance between the productive power and the price of the factors labour and energy as described in Chap. 3—labour: low productive power with a high factor cost share, energy: high productive power with a low factor cost share—as well as the problems of emissions and employment, the idea of transferring the principle of taxation according to economic performance from persons to the factors of production is obvious.

If, by means of taxes on the natural gift of energy and its use to reduce non-wage labour costs and the tax burden on employees' incomes, the prices of energy and labour were to some extent brought closer to their productive levels, cost differentials such as those shown in Fig. 3.7 in the direction of decreasing labour input and increasing energy input would weaken. Rationalization pressures would be alleviated. Social disparities within society would also be reduced, especially if the reform were to benefit primarily the 50% of German households that together own only 2.5% of Germany's private wealth. A generational conflict between young and old as a result of declining birth and death rates and a possible reversal of the age pyramid would also be avoided: The energy slaves would pay the pensions.

Investment in rational energy use techniques would be more profitable with gradually and predictably rising energy prices—especially if energy taxes were handled flexibly and buffered energy price fluctuations on the world market (After the crash in oil prices in the first half of the 1980s, shown in Fig. 3.1, many rational energy use projects were abandoned). Exploiting the potential of rational energy use would also reduce the discrepancy between the per capita CO_2 emissions of the industrialized countries and those of the developing and newly industrializing countries.

6.1.2 Border Adjustment Levies

The Commission of the European Community had proposed on 25 October 1991 that a combined energy-CO_2 tax be introduced EU-wide from 1 January 1992, with an initial rate of US$3 per barrel of oil equivalent, rising to US$10 per barrel by the year 2000. On 11 December 1991, the German government had welcomed this proposal. However, it was not implemented because vested interests pushed for its suspension until the US and Japan did the same.

Considering the national and international resistance to the concepts of ecological tax reforms discussed in the 1990s, the "Tranche I Taxation Study" prepared in preparation of the UN Framework Convention on Climate Change in Kyoto stated [216]: 1. that energy expenditure accounts for only a relatively small share of the gross domestic product of the OECD countries (between three and 11% on the basis of purchasing power parity, with the OECD average being 5.8%). Nevertheless, energy-intensive industries would lose competitiveness if all other things remained equal and other partners did not introduce appropriate energy-carbon taxes. That is why an expert group of the UN Framework Convention on Climate Change wrote in a study on the impact of a combined energy-carbon tax of US$100 per tonne of carbon that it is often proposed to minimise the competitiveness problem of energy-carbon taxation through border adjustment levies. These border adjustment levies are currently being discussed by the WHO and the OECD with regard to their legal requirements and practicability [217].

Border adjustment levies mean that exporters from an economic area with energy taxes would be refunded all taxes on the energy used to produce the exported goods; on the other hand, importers would have to pay taxes/duties on goods imported from an energy tax-free economic area, based on the energy used to produce and transport these goods. The manifold political and legal problems associated with this have not yet been resolved. But if the European Union introduced energy taxes and border adjustment levies it should be possible to handle conflicts with trading partners.[4] Perhaps the US administration, in office since early 2017, would even play into the Europeans' hands in this regard.

6.1.3 Energy Taxes Under Discussion

In discussions about energy taxes, corporate executives in particular turn themselves into advocates for the socially disadvantaged: Employees would be burdened in a socially unbearable way by increased fuel prices due to their car journeys to work. The same would apply to increased energy costs in the household. As understandable as the rejection of energy taxes by managers is, because they may reduce company profits and the bonus payments linked to them, on the whole social

[4] At least, in order to protect bananas from French overseas territories, the EU has imposed special tariffs on banana imports from the dollar area, despite WHO world trade agreements.

hardship can be avoided without additional bureaucracy by increasing the kilometre allowance for wage and income tax and the rent subsidy for housing benefit recipients. However, an increase in the cost of leisure transport, domestic energy waste and all energy-intensive goods and services would hit many. But as with average factor costs, the shares of total private expenditure (excluding transport) accounted for by energy costs are more likely to be in the single-digit percentage range than above it. Consumption cuts as a result of higher energy taxes are likely to be accepted by the population if it becomes clear that the reduction in non-wage labour costs, which account for almost half of labour costs in Germany, counter-financed by energy taxes, reduces the pressure to cut jobs and increases social security. In more detail [218], points to hardship avoidance and social fine-tuning in the introduction of energy taxes.

Nevertheless, energy taxes remain an irritant. With the current state of information, without knowing the mechanisms of hardship avoidance mentioned, lower-income sections of the population fear a loss of prosperity if energy services for mobility and living comfort become more expensive. Their concern, as well as the fears of top earners that they will no longer be able to increase their incomes far above average through energy-supported rationalisation measures, as they have been able to do since the end of the Cold War, is reminiscent of the rejection of slave liberation and the abolition of serfdom in the agrarian feudal age: a former slave or serf had to be paid the wages of a free man. Energy taxes are the wages that the energy slaves, who no longer toil at almost zero cost, deliver to the state and society.

But even if there were at least theoretical agreement in society on the sense and purpose of energy taxes, there are a number of high hurdles to be considered against their introduction in Europe. The political scientist Armingeon described them in 1995 [219]:

1. The material interests of the sectors, companies and employee groups directly affected.
2. The structure of Western European party systems, which are much more shaped by historical socio-cultural divisions than by the new contentious issues of modern societies.
3. The interconnectedness of policy decisions between the superimposed levels of the political system.
4. The international interdependence of nation states and the unanimity rule of the European Union.
5. The past imprint of a country's structures and policies. In particular, its tax administration has been shaped at enormous cost over many years for a particular tax system for which no one can foresee the effects of change.
6. The problems of translating political decisions into administrative action, which sometimes even achieves the opposite of the intended objective.

The upheavals since 2001 are beginning to dissolve the structures of Western European party systems. This, and the new contentious issues concerning energy, the environment and migration, are making themselves felt in elections and referendums. Developing new fiscal instruments and equipping state institutions to use them can help reduce social instability. But this may not be enough. In that case, it is worth looking over the rim of the biosphere.

6.2 Extraterrestrial Production

> Our civilisation is dependent on satellite-based technologies in many areas. In order for future generations to be able to use such technologies, near-Earth space must be understood as an infrastructure that must be preserved for use in international cooperation. (D. Hampf, L. Umbert, Th. Dekorsky, W. Riede [222])

> I dream that we humans will work even better together and go far out into space – to the Moon, Mars and beyond. I wish that humanity would soon realize that it is merely a matter of choice to set out on such adventures. (Alexander Gerst, German ESA astronaut [223])

Should humans find it too crowded on their blue planet in the face of population growth, civil wars, international conflicts and thermodynamic constraints, there would be the possibility of industrial expansion into near-Earth space. Princeton physics professor *Gerard K. O'Neill* has proposed a way to do this, using solar power satellites and extraterrestrial manufacturing facilities to supply the Earth with solar energy and colonize near-Earth space.

O'Neill's vision, which from 1974 [63] until O'Neill's death in 1992 met with strong response from people of many different professions, ages, and nationalities, is based on two key elements: a space shuttle-based transport system for reaching low Earth orbit and a primarily photovoltaic-powered heavy-lift transporter to the Moon using a version of the electromagnetic mass drivers outlined in Sect. 2.3.2 as a rocket engine. Its reaction mass would consist of shredded external tanks of the Space Shuttle successors. With lunar material shot by a mass driver installed on the Moon into capture facilities in space, and energy from the Sun, humans living first in habitats on orbits around the Lagrange libration point L5 at equal distance from Earth and Moon would build solar power satellites [220, 221] to provide Earth with electricity. Summaries of the plans for space industrialization can be found in [64–66, 92], among others.

Microwave generators on satellites in geostationary orbit would convert energy generated photovoltaically by solar panels or thermoelectrically by mirrors, gas heaters and turbines into microwaves with frequencies of 2–3 GHz. These would be transmitted by the satellites' transmitting antennas to terrestrial receiving antennas, which would convert them back into electrical energy. Satellites in geostationary orbit, which weigh 34,000–86,000 tonnes depending on their design, would feed 5000–10,000 MW per satellite into the Earth's energy supply network almost without interruption.

The diameters of the transmitting antenna of a satellite and the receiving antenna on Earth are about 1 and 10 km. Four to eleven times more solar energy is accessible to the satellite than to the sunniest areas of the Earth, and this energy is available to it almost continuously except for brief periods of shading by the Earth. On an annual average, the shadows reduce the energy yield by 1% of the amount that would be gained if the solar power satellite were continuously exposed to solar radiation. Conversely, the core shadow of a 10,000 MW satellite does not reach the Earth.

Since a satellite in geostationary orbit does not move relative to the earth, the sharply focused microwave beam can be directed close to the major energy consumers, thus avoiding transmission losses in long lines.[5] The microwaves in the 3 GHz range penetrate the atmosphere and clouds with only minor losses and are converted into electrical energy in the receiving antenna with an efficiency of 90%. So little waste heat is generated in the process that approaching the heat barrier with solar power satellites is slower than with all other power generators of comparable performance.

The political and economic implementation of O'Neill's ideas was called for by House Concurrent Resolution 451 of both houses of the 95th U.S. Congress, introduced by Representative Olin Teague on December 15, 1977. It concludes with the call:

> To assist in these efforts the Office of Technology Assessment specifically is requested to organize and manage a thorough study and analysis to determine the feasibility, potential consequences, advantages and disadvantages of developing as a national goal for the year 2000 the first manned structures in space for the conversion of solar energy and other extraterrestrial resources to the peaceable and practical use of human beings everwhere [66].

This resolution was referred to the Committee on Science and Technology.

After Ronald Reagan became President of the United States of America, his science advisor George Keyworth declared at one of the "Princeton Conferences on Space Manufacturing Facilities" that the President was convinced that the peaceful development of space should be a matter for private enterprise.

But manned spaceflight is expensive, and the private sector—except for a short-lived "Council on Power from Space" of leading space companies—did not get involved in O'Neill-type plans to industrialize near-Earth space. NASA's initial estimate of $10 million per space shuttle flight had risen to $180 million by 1985 after the first shuttle mission in 1981. After the *Challenger* and *Columbia* space shuttle disasters in 1986 and 2003, and a total of 135 launches of five different shuttle specimens, the US ended the space shuttle program with the last mission of *Atlantis* in 2011; the final cost per mission, including all retrofits, is estimated at US$500 million [225]. Since then, to transport its astronauts to the International Space Station (ISS), the US has been purchasing seats on Russian Soyuz spacecraft at a cost of US$51 million per passenger. The space shuttles could seat up to seven astronauts.

Measured in terms of human needs, the energy and material resources as well as the emission absorption capacity of space are virtually unlimited. Their economic

---

[5]The highest energy intensity of the microwave beam at its center is about half the intensity of sunlight. That is why the satellite solar power plants are completely useless as a weapon and would not fry birds flying through the microwave beam.

exploitation through extraterrestrial technologies still seems like science fiction to most contemporaries. But it should be technically feasible. In the executive summary of the study "Space-Based Solar Power As an Opportunity for Strategic Security" of the *National Space Security Office* from 2007, it is stated [224]:[6] "The threateningly growing energy and environmental problems are so serious that they require the consideration of all options for their solution. This includes rediscovering plans to build solar power satellites first developed in the U.S. nearly 40 years ago." The final recommendation is that "The study group recommends that the U.S. government soon embrace these plans, demonstrate their feasibility, and promote their implementation."

As with all large-scale projects, cost is the problem.

When assessing costs, it may perhaps serve as a benchmark that German electricity customers paid photovoltaic electricity producers around EUR 10 billion in feed-in tariffs via the Renewable Energy Sources Act in 2016, see Fig. 5.1 and the explanations. In return, they received an electrical energy quantity of 38.2 TWh from a PV capacity of around 40 GW according to Table 2.5. A 10,000 MW (=10 GW) solar power satellite would supply 87.6 TWh per year.

If, according to the plans of Boeing and other aerospace companies, all satellite components were transported from Earth to geostationary orbit with 60–100 heavy lift rocket launches per satellite, the cost of a 10,000 MW satellite would be US$10 to 15 billion, according to estimates by the "Solar Power Satellite Study Manager for Boeing", Woodcock [226]. More environmentally friendly, and according to the cost estimates in [66, 92] cheaper, would be O'Neill's way, which could have led to the settlement and industrial use of near-Earth space already at the end of the twentieth century.[7]

With O'Neill's death, the hope of overcoming the limits of our fragile biosphere seemed extinguished. But it is reviving. Before setting off on his second mission on the International Space Station (ISS), Alexander Gerst reports:

> ESA is working with NASA on the Orion spacecraft, which will fly far beyond low Earth orbit. As a next step, an international group of space agencies is preparing to go even deeper into space: a base camp outside the Earth's orbit, the 'Deep Space Gateway'. We have a chance to be part of this venture. [223]

Perhaps, during the present years of global economic growth, humanity will begin to open up new economic and living spaces beyond the Earth, instead of limiting its urge for change and discovery to an overpopulated Earth with its thermodynamic

[6] Translation R.K.

[7] According to [66], the investment costs of US$60 billion until the first energy delivery of the satellite power plants to Earth should be amortized in about 20 years. It is true that this calculation was based on the costs of US$10 million per shuttle flight, which were initially estimated too low by NASA. This contrasts with an assumed interest rate of 10% on the investments, which is significantly higher than the various falling interest rates of the twenty-first century.

growth limits, earthquakes, tsunamis, hurricanes and, beneath Yellowstone Park and the Phlegraean Fields, supervolcanoes with high extinction potential.

6.3 Outlook

The report of the Club of Rome *The Limits to Growth* from 1972 [95] and the first oil price shock of 1973 triggered worldwide reflection on whether the "growth dogma", which is still valid in important schools of economic thought today, should not be abandoned or modified. Even then, the limits to energy-driven economic growth [24, 227] outlined in Chap. 1 and drawn by entropy production were well known. Many of those who did not and do not turn a blind eye to these limits followed and still follow the conclusions of [95]: industrial production on earth must be transferred from the (post-war) state of exponential growth to a state of equilibrium. The *steady state economy* [161] is in the long run the only option for the "spaceship earth" [228]. Chapter 4 describes the necessary behavioural changes from today's perspective.

In 1975, under the impression of my Colombia experiences reported in Chap. 5, I proposed to the economist and theologian Wilhelm Dreier of the Chair of Christian Social Science at the University of Würzburg a joint interdisciplinary seminar on economic growth, energy and the environment. As part of the research focus "Endangered Future" of the Faculty of Theology, it was very well received by students and teachers from the natural sciences and the humanities in the second half of the 1970s. Results were published in [211] and elsewhere. Of course, it was also pointed out that the thermodynamic growth constraints with their outermost limit, the heat wall, apply to terrestrial industrial production, and that thermodynamics does not stand in the way of industrial expansion into near-earth space as described in Sect. 6.2. One objection raised to space industrialization was, "But then human nature need not change." This was coupled with the hope that the change would make people better Christians. To this, the most experienced Würzburg representative of Catholic moral theology, Professor Heinz Fleckenstein, replied: "According to a long-established principle of Catholic moral teaching, the technical solution to a problem is always preferable to the attempt to change human nature."

The economic historian Professor Robert Heilbroner is deeply pessimistic about the future of a humanity that will have to make do with the earth's natural resources and disposal mechanisms in a stationary society and will sooner or later have to make the transition from the dynamic growth economy with relatively high social mobility[8] to a stationary state. If only an authoritarian, or possibly only a revolutionary regime, will be able to accomplish the vast social reorganization necessary to avoid catastrophe, he fears for a stationary society restrictions on the pursuit of scientific knowledge, the enjoyment of intellectual heresy, and the freedom to arrange one's life as one sees fit. In his book *An inquiry into the Human Prospect*

[8] From working-class child to Chancellor of the Federal Republic of Germany.

[229] he concludes: "If then, by the question, Is there hope for man?" we ask whether it is possible to meet the challenges of the future without the payment of a fearful price, the answer must be: No, there is no such hope.

In the stationary societies of the past with peasant servitude, craftsmen's guilds, castes, noble feudal lords, court officials and (priest) kings, individuals generally remained confined to their social class. Clerical occupations offered certain opportunities for advancement. It is true that in imperial China, in the regular civil service examinations since the eleventh century, the lowliest peasant could in principle rise to the highest minister of the empire. Those who had already risen in rank and failed an examination had to be relegated again. But could such a meritocratic system be established in a stationary world society of 10 billion people, with memories of the good life in the dynamic industrial societies of the decades after World War II, without corruption, authoritarian structures and suppression of intellectual freedom?[9]

The social and technical creativity of the generations of the twenty-first century is confronted with an unprecedented challenge. Meeting it is likely to be all the more successful the more clearly the economic consequences of the thermodynamic laws are seen, and the more powerfully the Golden Rule—Judeo-Christian version: Love thy neighbor as thyself—governs human coexistence. And certainly it would do no harm if the boldest and best men and women followed the L5 Society's call in the 1980s: "If you love the Earth, leave it!"

However, if the laws of nature and the ethical recommendations of the great religions and wisdom teachings do not soon become the basis of global economic and political action, and the striving for awakening and the opening up of new worlds is limited to *virtual reality,* what the ancient Greeks had already experienced and summed up will probably take place in the global upheaval: Learning through suffering. We still have the choice.

[9] It is still too early to see the decisions of the PRC's National People's Congress in March 2018 and the resulting power grab by President Xi Jinping, as well as the rise of authoritarian politicians to the top of other states, as harbingers of future developments to maintain domestic order around the world.

Appendices

A.1 Entropy, Environment, Information

A.1.1 Entropy Production and Emissions

In a non-equilibrium system with volume V, entropy is produced per unit time d_iS/dt. The *entropy production density* $\sigma_S(\mathbf{r}, t)$ produced inside the volume at the location $\mathbf{r}$ at time t is defined by the identity

$$\frac{d_i S}{dt} \equiv \int_V \sigma_S(\mathbf{r}, t) dV. \tag{A.1}$$

The volume V can also be so tiny that the right-hand side of Eq. (A.1) is simply equal to $\sigma_S \cdot V$. Thus it follows from the Second Law, i.e. from $d_iS/dt > 0$ according to Eq. (1.3), that also the entropy production density σ_S itself is everywhere and at any time unavoidable, i.e. greater than zero:

$$\sigma_S(\mathbf{r}, t) > 0. \tag{A.2}$$

The entropy production density consists of a fraction $\sigma_{S,\ chem}$, caused by chemical reactions, and a fraction $\sigma_{S,\ dis}$, associated with heat flows and particle flows. Thus, Eq. (A.2) becomes

$$\sigma_S(\mathbf{r}, t) = \sigma_{S,chem} + \sigma_{S,dis} > 0. \tag{A.3}$$

Here $\sigma_{S,\ chem}$ is given by *scalar* generalized currents and forces, while in $\sigma_{S,\ dis}$ there are the *vectorial* currents and forces (or their densities) shown in Eq. (A.4) [230]. These two terms cannot interfere with each other. Therefore, because of Eq. (A.3), each of the two terms must be positive by itself, i.e.: $\sigma_{S,\ chem} > 0$ and $\sigma_{S,\ dis} > 0$.

For the analysis of entropy production processes such as the combustion of a certain amount of coal, oil or gas, one can first consider the process of chemical transformation in which $\sigma_{S,\ chem}$ is produced, and *after* which there are N different kinds of particles in the combustion plant and a certain environment (chimney),

R. Kümmel et al., *Energy, entropy, creativity*,
https://doi.org/10.1007/978-3-662-65778-2

which are numbered by k, and which then spread throughout the system according to the generalized forces acting on them.[1] In their excellent book *Grundlagen der Thermodynamik*, Kluge and Neugebauer [231] have shown that the associated "dissipative" entropy production density (called in [2]) $\sigma_{S,\ dis}$ is given by[2]

$$\sigma_{S,dis}(\mathbf{r},t) = \mathbf{j}_Q \nabla \frac{1}{T} + \sum_{k=1}^{N} \mathbf{j}_k \left[-\nabla \frac{\mu_k}{T} + \frac{\mathbf{f}_k}{T} \right] > 0. \tag{A.4}$$

Here $\mathbf{j}_Q(\mathbf{r},t)$ is the heat flux density and $\mathbf{j}_k(\mathbf{r},t)$ denotes the diffusion flux density of the particle species k. $\mathbf{j}_Q$ is driven by the gradient ∇ of the absolute temperature $T(\mathbf{r},t)$ and $\mathbf{j}_k(\mathbf{r},t)$ is driven by gradients of the chemical potential $\mu_k(\mathbf{r},t)$ and the temperature as well as by specific external forces $\mathbf{f}_k(\mathbf{r},t)$. The heat flux densities and partial flux densities are the carriers of the emissions referred to in Eq. (1.3).

N. Georgescu-Roegen, who has the great merit of having pointed out the importance of entropy for the economy [232], believed to have discovered a Fourth Law of thermodynamics: that of the dissipation of matter. This led to heated discussions for a while, until it was clarified that the second term on the right-hand side of (A.4) covers what Georgescu-Roegen had in mind [233].

A.1.2 Greenhouse Effects

Natural Greenhouse Effect

The simplest quantitative description of the natural greenhouse effect considers the Earth in radiative equilibrium with the Sun. The flux of solar radiation power at the edge of the Earth's atmosphere, the *solar constant S*, *is* measured by balloon, rocket and satellite experiments to be

$$S = 1367\ \ \mathrm{W/m^2}. \tag{A.5}$$

About 30% of the solar irradiance S, the so-called albedo α, is reflected back into space. Thus, at the edge of the Earth's atmosphere, a radiation power flux of

$$S(1-\alpha)/4 = 239\ \ \mathrm{W/m^2} \tag{A.6}$$

is absorbed. The factor 1/4 is the quotient of the cross-sectional area and the surface area of the earth. (Here the earth is approximated by a sphere with radius R, so that its surface has the size $A_E \approx 4\pi R^2$ and its cross-section has the area πR^2.) The product of

[1] Stepwise analyses of combustion processes are also carried out for the calculation of the energy (=exergy) which can be gained in such processes and which can be used for work [49], see also Eq. (A.26).

[2] The detailed vector-analytical calculations for the derivation of Eq. (A.4) taken from [231] are briefly sketched in the appendix to the chapter "Entropy" of [2].

$A_E = 510 \times 10^6 \ \text{km}^2$ with 239 W/m² results in the solar radiation power *absorbed by* the earth to be

$$P_{solar} = 1.2 \cdot 10^{17} \ \text{W}. \tag{A.7}$$

The solar radiation power *impinging* on the upper edge of the atmosphere, i.e. not reduced by the albedo, is then

$$P_{top} = 1.7 \cdot 10^{17} \ \text{W}. \tag{A.8}$$

The spectral range of the radiant power absorbed by the biosphere according to (A.6) and (A.7) extends from 0.2 to 2 μm. This power is radiated back into space in the infrared spectral range between about 5 and 30 μm. The effective temperature T_{eff} required for this is calculated according to the Stefan-Boltzmann law (in the approximation for "black bodies") from the equation

$$S(1-\alpha)/4 = \sigma T_{eff}^4 \tag{A.9}$$

using (A.6) to

$$T_{eff} = 255\text{K} = -18 \ ^\circ\text{C} \tag{A.10}$$

Here $\sigma = 5.67 \cdot 10^{-8} \text{W/m}^2\text{K}^4$ is the Stefan-Boltzmann constant.

However, the solid surface of the earth has an average temperature of +15 °C = 288 K. At this temperature, the Stefan-Boltzmann law yields a radiated heat flux density of 390 W/m². The difference of 151 W/m² between the heat flux densities radiated from the Earth's surface on the one hand and from the upper edge of the Earth's atmosphere on the other hand is "captured", as it were, by the infrared-absorbing trace gases of the Earth's atmosphere and partly radiated upwards and partly downwards again. These gases thus surround the Earth with a warming mantle of radiation that sustains our life. They play roughly the same role as the glass roof and walls of a greenhouse: visible sunlight passes through the glass almost unhindered and is absorbed by the soil and the plants in the greenhouse, which are thus warmed. The absorbed heat is re-radiated by the soil and plants in the infrared range. This radiation is in turn absorbed by the surrounding glass surfaces and then partly radiated inwards and partly outwards. The re-radiated heat raises the temperature in the greenhouse above the outside temperature. (The prevention of convection by the glass roof contributes additionally to the heating).

The infrared active trace gases (greenhouse gases) in the atmosphere, their concentrations in "parts per million" (ppm), and their contributions to temperature increase above −18 °C are: Water Vapor, H_2O (between 2 ppm up to 3×10^4 ppm, 20.6 °C), Carbon Dioxide, CO_2 (pre-industrial: 280 ppm, 7 °C), (ground level) Ozone, O_3 (0.03 ppm, 2.4 °C), Nitric Oxide, N_2O (0.3 ppm, 1.4 °C), Methane, CH_4 (1.7 ppm, 0.8 °C) and others (0.6 °C).

Anthropogenic Greenhouse Effect

If the concentrations of infrared-active trace gases in the atmosphere increase due to human activities, the warming radiation mantle with which they surround the Earth becomes denser, and all the energy radiated by the Sun can only be radiated back into space as heat by increasing the surface temperature of the Earth.

The essential point about the anthropogenic greenhouse effect (AGE) is that the absorption maxima of the greenhouse gases (other than water vapour) for thermal radiation lie in the "windows" of the infrared absorption spectrum of water vapour (H_2O), the strongest of all greenhouse gases. These windows, through which much heat radiation escaped into space in pre-industrial times, are being closed ever more tightly as the concentration of greenhouse gases released by humans increases.

Energetically, the most important open windows of atmospheric water vapor are between 7 and 13 μm—where the infrared radiation of the Earth's surface reaches a maximum—and between 13 and 18 μm—where the infrared radiation is only partially absorbed by the water vapor. In the central region of the open window between 8 and 12 μm the quasi-continuous absorption of thermal radiation by water vapour plays an important role. It increases with the square of the water vapor pressure. Because of the exponential increase of the partial pressure of H_2O with temperature, there is a strong feedback between temperature increase and infrared absorption by water vapor. Therefore, increasing CO_2 concentrations make a large contribution to the AGE (although the two main absorption bands of CO_2 at 15 and 4.3 μm are already mostly saturated): Additional CO_2 causes only a small temperature increase δT; but the partial pressure of atmospheric H_2O grows exponentially with δT, and infrared absorption by water vapor grows quadratically with its pressure. All together, the AGE caused by additional CO_2 grows with the logarithm of CO_2 concentration, so that for every doubling of atmospheric CO_2 concentration, the Earth's surface temperature increases by the same amount (of about 2.5 °C).

A.1.3 Entropy and Information

The canonical sum of states of a system of many interacting particles in contact with a heat bath of absolute temperature T is given, with $\beta = 1/kT$,

$$Z = \sum_R e^{-\beta E_R}. \tag{A.11}$$

E_R is the energy of *a* many-particle state indexed by R, and is summed over all many-particle states R *of* the system. One of the possible many-particle states (in the simplest case of an ideal gas) is shown in Fig. 1.1. The probability that the system is encountered in a certain state R during a measurement is

$$P_R = \frac{e^{-\beta E_R}}{Z}. \tag{A.12}$$

The mean energy of the many-particle system is given by the expectation value

$$\overline{E} \equiv \sum_R P_R E_R. \tag{A.13}$$

Textbooks of thermodynamics and statistics like [162] show that the entropy defined in Sect. 1.1 by $S = k_B \ln \Omega$ also can be calculated from

$$S = k\left[\ln Z + \beta \overline{E}\right] \tag{A.14}$$

If we insert $\overline{E}$ from Eq. (A.13) and note that

$$\sum_R P_R = 1 \tag{A.15}$$

you get after a short calculation

$$S = -k \sum_R P_R \ln P_R. \tag{A.16}$$

Here the constant k is equal to the Boltzmann constant k_B and $\ln P$ is the natural logarithm of P.

In information theory, *Shannon*, starting from Boltzmann's H-theorem (see, e.g., [162, pp. 624–626]), defined the entropy of a discrete random variable X with possible values $\{x_1, \ldots x_i \ldots x_n\}$ as the expected value of the information content of X:

$$H = -k \sum_i P(x_i) \ln P(x_i). \tag{A.17}$$

Here is $P(x_i)$, with

$$\sum_i P(x_i) = 1, \tag{A.18}$$

is the probability that the random variable X takes the value x_i. Shannon entropy is a measure of uncertainty/randomness (This uncertainty vanishes by observing (A.18) if an event $x_{i\,=\,e}$ occurs with certainty, so that $P(x_e) = 1$ and $\ln P(x_e) = 0$).

The fixation of the constant k corresponds to a fixation of the information unit. If one chooses $k = 1/\ln 2$, then, because of $\log_2 P = \ln P/\ln 2$ in the dual system with the information unit bit, one obtains the Shannon entropy as

$$H = -\sum_i P(x_i) \log_2 P(x_i). \tag{A.19}$$

Examples:

- $X = \{0, 1\}$ is a message in which the two characters 0 and 1 occur with the same probability of 1/2 each. Then the information content of this message is $H = -(1/2)\log_2(1/2) - (1/2)\log_2(1/2) = -\log_2(1/2) = 1$.
- $X = \{x_1, x_2, x_3, x_4, x_5, x_6, x_7, x_8, x_9, x_{10}\}$ is a message of ten characters. All characters may occur again with the same probability, so that $P(x_i) = 1/10$ is. Then the information content of this message is $H = -10\frac{1}{10}\log_2(1/10) = \log_2(10) \approx 3.322$.

The relation between entropy and information is discussed in more detail in [28]; for Shannon entropy see there the Sects. 1.6 and 3.3.

A.2 Energy and Exergy

A.2.1 Energy Units

1 joule [J] = 1 watt-second [Ws]
1 million tonnes of coal equivalent (tSKE) 1 MtSKE = 29.3 PJ
1 million tonnes of oil equivalent (toe) = 1 toe = 41.9 PJ
1 tonne of oil = 7.3 barrels of oil (1 barrel = 159 litres)
historical unit: calorie (cal). 1 cal = 4.19 J

Power units
1 Watt (W) = 1 Joule per second (J/s)
1 HP (1 horsepower) = 0.7355 kW

A.2.2 Energy Quantity and Quality

For energy services, in addition to the quantities of energy measured in joules (J), kilowatt hours (kWh) or tons of hard coal units (tSKE), their quality is also important. The energy quality is determined by the *exergy* contained in an energy quantity.

Consider two systems of different atoms and molecules that are initially isolated from each other. Their respective internal energies, measured at constant volumes, are U_1' and U_1'', and the total initial internal energy is $U_1 = U_1' + U_1''$. Then the systems are brought into contact and interaction with each other. During the chemical reactions, the electrons and atoms are rearranged. When the total system has found a new equilibrium, its internal energy is U_2.

Most chemical reactions take place at constant pressure p and not at constant volume. If the total volume of the two isolated systems is $V_1 = V_1' + V_1''$ at the beginning and the total volume is V_2 after the reaction, the work $p(V_2 - V_1)$ is exchanged between the total system and the environment. The environment is in many cases the atmosphere.

The difference $U_2 - U_1$ between the final and initial internal energies is then given by

$$U_2 - U_1 = -p(V_2 - V_1) - Q_{12}. \tag{A.20}$$

Here Q_{12} is the heat that is released into the environment during the reaction. The chemical reaction is called exothermic (endothermic) when $Q_{12} > 0$ ($Q_{12} < 0$).

If the heat balance of a chemical reaction is of interest, it is useful to introduce the thermodynamic state function H called *enthalpy*. It is defined as

$$H \equiv U + pV. \tag{A.21}$$

Thus, Eq. (A.20) can be written as

$$H_1 - H_2 = Q_{12}. \tag{A.22}$$

If both initial and final states of the chemical reaction are liquid or solid, the difference between enthalpy H and internal energy U is small. However, if gaseous states also play a role, as is the case with the combustion of fossil fuels, the heat generated must be calculated from Eq. (A.22).

When determining the energy *quantities* contained in fossil fuels, all reaction partners are transferred from an initial state before the heat-generating reaction to a reference state after the reaction. The initial state is characterized by the enthalpy H_1 and the final state by the enthalpy H_2. Then the heat released is given by $Q_{12} = H_1 - H_2$.

The heat quantities resulting from enthalpy differences indicate the *amounts of energy* contained in the fuels.

Normally, the temperature of the reference state 25 °C ≈ 298 K is selected. The pressure is often chosen to be that of the natural environment. The reference state of each chemical element is the most stable naturally occurring compound of that element. For carbon and hydrogen, these compounds are CO_2 and gaseous or liquid water (H_2O). One measures the amounts of energy from coal, oil, gas, and biomass in a controlled combustion process. In this process, fuel and air with the same initial temperature T_0 are brought into a reaction chamber (calorimeter), and the combustion products must be cooled down to exactly the temperature T_0. The heat that is then transferred to the reaction chamber is called the calorimeter. The heat that then leaves the reaction chamber divided by the amount of fuel is the specific calorific value of the fuel. There is an upper and a lower heating value. The two values differ by the heat of vaporization of H_2O. Table 2.1 gives average heating values.

Energy *quantities*, measured in enthalpy units such as joules (or kilowatt-hours or tons of coal or oil equivalents), are not sufficient to characterize the usefulness of an energy source. Energy *quality* is also important. According to Karlsson [234] and van Gool [235], the quality of energy is defined as

Quality = exergy/enthalpy.

Exergy is that part of a quantity of energy which can be completely converted into physical work. As explained in Sect. 1.1.3, it is supplemented by the useless anergy in the law of conservation of energy.

Examples of Exergy

1. The kinetic energy of a mass m with velocity $\mathbf{v}$ is 100% exergy. The same applies to the potential energy of a mass in a gravitational field.
2. Electrical energy is 100% exergy.
3. The chemical energy stored in coal, oil and gas is in principle and ideally 100% exergy. The same applies to the energy obtained from mass conversion according to $E = mc^2$. This is so because, in principle, the combustion of fossil fuels and the extraction of energy from mass conversion can take place at very high absolute temperatures T. Then the exergy E_X of the heat Q driving a Carnot machine operating between a reservoir of absolute temperature T and the environment of absolute temperature T_0 is given by the Carnot efficiency:

$$E_X = Q(1 - T_0/T) \rightarrow Q \text{ for } T \gg T_0.$$

   A Carnot machine is an idealized heat engine that reversibly, i.e. infinitely slowly, cycles. The efficiencies of real heat engines are smaller than $(1 - T_0/T)$.
4. Also the solar radiation consists in principle of 100% exergy. Karlsson [234] has shown that the quality of quasi-monochromatic, incoherent radiation in the frequency range between ω and $\omega + \mathrm{d}\,\omega$, falling perpendicularly on the surface of a black body of temperature T_0, is given by

$$\frac{\text{Exergy}}{\text{Enthalpy}} = 1 - T_0/T + [\exp(\hbar\omega/k_BT) - 1] \times \frac{k_BT_0}{\hbar\omega} \ln \frac{[1 - \exp(-\hbar\omega/k_BT)]}{[1 - \exp(-\hbar\omega/k_BT_0)]}. \tag{A.23}$$

   Here is

$$T = \hbar\omega / \left\{k_B \ln\left[\left(2\hbar\omega^3/c^2P_E\right) + 1\right]\right\} \tag{A.24}$$

   the equivalent temperature of a black body radiating per unit area the power $P_E(\omega)\,\mathrm{d}\,\omega$ in the frequency range between ω and $\omega + \mathrm{d}\,\omega$. If T is the effective surface temperature of the Sun of 5777 K, and is $T_0 = 288$ K, which is the mean surface temperature of the Earth, then the quality of the equivalent blackbody radiation is very close to 1 [234].
5. A many-particle system with internal energy U, entropy S, and volume V that is *not* in equilibrium with its environment of temperature T_0, in which there is

pressure p_0 and with which it can exchange heat and work—but not matter—, has exergy [49, 50]

$$E_X = (U - U_0) + p_0(V - V_0) - T_0(S - S_0). \tag{A.25}$$

Here U_0, V_0, and S_0 are the internal energy, volume, and entropy of the system when it has reached equilibrium with its environment.

6. Into a thermodynamic system with volume V and pressure p, a steady stream of mass m with kinetic energy $m\mathbf{v}^2/2$ enters at one point; at another point at the same height z above a reference point in the gravitational field with acceleration g, it exits again. This system contains with respect to the environment with U_0, V_0, p_0 and S_0 the exergy [49, 50]

$$E_X = (U + pV - U_0 - p_0V_0) - T_0(S - S_0) + m\mathbf{v}^2/2 + mgz.$$

7. A combustion process produces a many-particle system of internal energy U and entropy S, whose volume is V. The system is not in equilibrium with its environment, in which the temperature T_0 and the pressure p_0 prevail and with which heat, work and matter can be exchanged. The combustion products are more concentrated in the combustion chamber than in the surroundings. In principle, work can be extracted from their diffusion into the environment. This work potential contributes to the exergy of the system. To calculate this exergy component, it is assumed that immediately after the combustion process, the system components are mixed and are already in thermal and mechanical equilibrium with the environment, but have not yet left the combustion chamber. Then they leave the combustion chamber and diffuse. If the system consists of N different particle species i, with n_i = number of particles of species i, and if μ_{i0} and μ_{id} are the chemical potentials of the partial echcomponent i in thermal and mechanical equilibrium before and after diffusion, then the exergy content of the system (neglecting the kinetic and potential energies) is [49]

$$E_X = (U - U_0) + p_0(V - V_0) - T_0(S - S_0) + \sum_{i=1}^{N} n_i(\mu_{i0} - \mu_{id}). \tag{A.26}$$

A.3 Aggregation

Assumptions

The monetary valuation of a good or service is on average the higher (irrespective of short-term fluctuations) the more physical work and information processing is required to produce the good or service.

The monetary valuation of work performance and information processing is usually different for different components of value added. But the mean value taken over all components of value creation remains constant to a good approximation during short time intervals. Thereby, a short time interval is smaller than the times characteristic for innovation diffusion.

The same applies to the monetary valuation of a capital good according to its capacity to perform physical work and process information.

A.3.1 Value Added

The kilowatt-hour (kWh) is chosen as the unit of energy input to produce a good or service, and the kilobit (kB) is the unit of information processing. (One kilobit is 1000 yes-no decisions, represented, for example, by switching electrical currents on and off). Unit sizes do not matter if dimensionless variables normalized to a base year are used for value added and factors of production.

Time series of value added in physical units are not reported by any statistics. Therefore, for econometric practice, they have to be mapped to the inflation-adjusted monetary time series published by the national accounts. For this purpose, a fictitious monetary unit is introduced as a placeholder for any real existing currency. We call this unit the "mark". We link the monetary measure of value added to the physical measure in five steps.

1. The monetary value added $Y_{mon}(t)$ of the economic system at time t, measured in marks, is divided into M ($\gg$1) parts $Y_{i,\ mon}$, all of which have the same monetary value μ marks: $Y_{mon} = \sum_{i=1}^{M} Y_{i,mon} = M\mu$ Mark. (Temporal changes in $Y_{mon}(t)$ result in corresponding temporal changes in $M(t)$; μ is constant and Mark $=Y_{mon}(t)/M(t)\mu$.)
2. The physical work done to produce $Y_{i,\ mon}$ is measured by the energy input required. This includes energy conversion losses due to friction and thermodynamic efficiency constraints. We define:
   $W_i \equiv$ Number of kilowatt hours of primary energy consumed in the production of. $Y_{i,\ mon}$
   $V_i \equiv$ Number of kilobits processed in the generation of $Y_{i,\ mon}$.
   Mechanized standards can be agreed upon for the measurements of W_i and V_i for all goods and services sold on the market, because these can in principle be regarded as produced by standardized routine procedures (The error in applying these measurement rules to the non-mechanised part of agriculture is smaller than the total contribution of agriculture to GDP).
3. We define the physical creation of value Y_{phys} of the economic system as

$$Y_{phys} = \sum_{i=1}^{M} Y_{i,phys} \equiv \sum_{i=1}^{M} W_i \cdot kWh \cdot V_i \cdot kB. \tag{A.27}$$

According to the definitions in step 1, $Y_{i,\ phys}$ has the monetary value μ Mark.

The unit of physical value added is called ENIN, as an abbreviation of "ENergy and Information". It is defined as the mean value

$$ENIN \equiv \frac{1}{M} \sum_{i=1}^{M} W_i \cdot V_i \cdot kWh \cdot kB = \zeta \cdot kWh \cdot kB, \quad (A.28)$$

where

$$\zeta \equiv \frac{1}{M} \sum_{i=1}^{M} W_i \cdot V_i \quad (A.29)$$

is such that

$$Y_{phys} = M \cdot ENIN. \quad (A.30)$$

The equivalence factor ζ changes with time t if the monetary valuation of work performace and information processing changes such that the numbers $W_i \cdot V_i$, corresponding to constant μ, *and* the right-hand side of Eq. (A.29) become time-dependent.

4. From the above equations, the relationship between monetary and physical value added is as follows

$$\frac{Y_{mon}(t)}{Y_{phys}(t)} = \frac{M\mu \cdot Mark}{M \cdot ENIN} = \frac{\mu}{\zeta} \cdot \frac{Mark}{kWh \cdot kB}. \quad (A.31)$$

As long as ζ is constant, there is proportionality between Y_{phys} and Y_{mon}. Then the monetary, inflation-adjusted time series of value added are proportional to the technological time series physically aggregated in ENIN.

5. If one works, as effectively done in Sect. 3.3, with dimensionless variables normalised to their magnitudes in a base year t_0, then the dimensionless value added at time t

$$y(t) \equiv Y(t)/Y_0, \quad (A.32)$$

with $Y_0 = Y(t_0)$. Consequently, the dimensionless physical value added at time t

$$y_{phys}(t) \equiv \frac{Y_{phys}(t)}{Y_{phys}(t_0)}. \quad (A.33)$$

Because of Eq. (A.31), the monetary value added in the base year is t_0

$$Y_{mon}(t_0) = \frac{\mu}{\zeta_0} \cdot \frac{Mark}{kWh \cdot kB} Y_{phys}(t_0), \qquad \zeta_0 \equiv \zeta(t_0). \quad (A.34)$$

Combining Eqs. (A.31), (A.33), and (A.34), it follows that the dimensionless time series of monetary value added equals the dimensionless time series of physically aggregated time series of value added multiplied by the factor $\frac{\zeta_0}{\zeta}$:

$$y_{mon}(t) \equiv \frac{Y_{mon}(t)}{Y_{mon}(t_0)} = \frac{\zeta_0}{\zeta} \frac{Y_{phys}(t)}{Y_{phys}(t_0)} \equiv \frac{\zeta_0}{\zeta} y_{phys}(t). \quad \text{(A.35)}$$

As long as the equivalence factor ζ can be regarded as time-independent and equal to ζ_0, the dimensionless monetary time series is equal to the dimensionless physical time series. If the equivalence factor ζ becomes time-dependent, the parameter $Y_0(t)$ of the LinEx function (3.42) differs from its value $Y_0(t_0)$ in the base year t_0.

A.3.2 Capital

The capital stock of an industrial economic system consists of all the energy conversion equipment and information processors, together with all the buildings and installations necessary to protect and operate them.

We relate the time series of the capital stock of an economic system published by the national accounts to the capacity of this capital stock to perform work and process information. For the aggregation of capital into physical units, we proceed in five steps, as in the case of value added.

1. The capital stock measured in monetary terms (in inflation-adjusted marks), K_{mon}, is divided into N ($\gg$1) units $K_{i,\ mon}$, all of which have the same monetary value ν marks: $K_{mon} = \sum_{i=1}^{N} K_{i,mon} = N\nu$ marks. (Temporal changes in $K_{mon}(t)$ result in corresponding temporal changes $N(t)$; ν is constant and Mark $=K_{mon}/N\nu$.)
2. The ability to perform work per unit time is measured in kilowatts (kW), and the ability to process information per unit time is measured in kilobits per second (kB/s). We define:
   S_i = number of kilowatts performed by the fully utilized capital good i with monetary value. $K_{i,\ mon}$
   T_i = number of kilobits per second that capital good i with monetary value $K_{i,\ mon}$ processes.
   The S_i is taken from the machine descriptions, and the T_i is given by the number of switching processes per time unit that allow or block the energy flows in the fully loaded machines.
3. We define the physical capital stock K_{phys} of the economic system as.

$$K_{phys} = \sum_{i=1}^{N} K_{i,phys} \equiv \sum_{i=1}^{N} S_i \cdot kW \cdot T_i \cdot kB/s. \quad \text{(A.36)}$$

According to the definitions in step 1, $K_{i,\ phys}$ has the monetary value ν Mark. The unit of physical capital is called *ATON*—abbreviation for AuTomatiON. It is defined as the mean value

$$ATON \equiv \frac{1}{N} \sum_{i=1}^{N} S_i \cdot T_i \cdot kW \cdot kB/s = \kappa \cdot kW \cdot kB/s, \quad \text{(A.37)}$$

with

$$\kappa \equiv \frac{1}{N} \sum_{i=1}^{N} S_i \cdot T_i, \quad \text{(A.38)}$$

so that

$$K_{phys} = N \cdot ATON. \quad \text{(A.39)}$$

The equilibrium factor κ changes with time t if the monetary valuations of the abilities to perform work and process information change in such a way that the numbers $S_i \cdot T_i$, corresponding to constant ν, *and* the right-hand side of Eq. (A.38) become time-dependent.

4. The relationship between monetary and physical capital results from the above equations to

$$\frac{K_{mon}(t)}{K_{phys}(t)} = \frac{N\nu \cdot Mark}{N \cdot ATON} = \frac{\nu}{\kappa} \cdot \frac{Mark}{kW \cdot kB/s}. \quad \text{(A.40)}$$

If κ is constant, there is proportionality between K_{phys} and K_{mon}. Then the monetary, inflation-adjusted time series of the capital stock are proportional to the technological time series physically aggregated in ATON.

5. If one works, as effectively done in Sect. 3.3, with dimensionless variables normalized to their magnitudes in a base year t_0, the dimensionless physical capital stock is

$$k_{phys}(t) \equiv \frac{K_{phys}(t)}{K_{phys}(t_0)}. \quad \text{(A.41)}$$

Because of Eq. (A.40) the monetary capital stock is at time t_0

$$K_{mon}(t_0) = \frac{\nu}{\kappa_0} \cdot \frac{Mark}{kW \cdot kB/s} K_{phys}(t_0), \qquad \kappa_0 \equiv \kappa(t_0). \tag{A.42}$$

Combining Eqs. (A.40), (A.41), and (A.42), it follows that the dimensionless time series for the monetarily aggregated capital stock is equal to the dimensionless time series for the physically aggregated capital stock multiplied by: $\frac{\kappa_0}{\kappa}$

$$k_{mon}(t) \equiv \frac{K_{mon}(t)}{K_{mon}(t_0)} = \frac{\kappa_0}{\kappa} \frac{K_{phys}(t)}{K_{phys}(t_0)} \equiv \frac{\kappa_0}{\kappa} k_{phys}(t). \tag{A.43}$$

As long as the equivalence factor κ can be regarded as time-independent and equal to κ_0, the dimensionless monetary time series is equal to the dimensionless physical time series. If the equivalence factor κ becomes time dependent, this contributes to the time dependence of the technology parameter a in the LinEx function (3.42).

A.3.3 Labour and Energy

The human contribution to value creation consists of routine work and creative work. The former, labour, is measured physically, in terms of hours worked per year (or, less precisely, in terms of the total number of people employed per year). The latter consists of the input of non-physically measurable ideas, inventions, and value decisions into the production process; our theory quantifies their contribution *ex post* as the operation of *creativity* from changes in technology parameters over time. In part, energy use, especially in information technology, can also be seen as a proxy variable for the educational level of the workforce.

The production factor energy is measured quantitatively by the enthalpy values of the energy carriers used and qualitatively by their exergy contents. Appendix A.2 deals with this in more detail.

A.4 Past and Future

A.4.1 Early Stage of Industrialisation

We call an economy "early industrialized" compared to that of a highly industrialized country discussed in Sect. 3.3.5 if its use of capital and energy per capita is much smaller than in the highly industrialized system.

Let us consider an early industrialized economic system in whose capital stock, in addition to machines (powered by fossil and solar energies), tools handled by human muscle power as well as farm animals, their stables and the implements combined with them play a role.

Then, if $\beta = 1 - \alpha - \gamma$ is used, the asymptotic boundary conditions apply

$$\alpha \rightarrow 0, \quad \text{provided that} \quad \frac{L/L_0}{K/K_0} \rightarrow 0, \quad \frac{E/E_0}{K/K_0} \rightarrow 0, \tag{A.44}$$

$$\gamma \rightarrow 0, \quad \text{provided that} \quad \frac{L}{L_0} \rightarrow \frac{L_{max}}{L_0} \equiv c\frac{K_{min}}{K_0}. \tag{A.45}$$

Equation (A.44) follows again from the law of diminishing returns. Equation (A.45) describes the approximation to a state which we call the "limit state of manufacture". In this state, the use of human labor is maximal, $L = L_{max}$, compared to a state with more machines of the above type, and the capital stock is minimal; c is the second free technology parameter.

Simplest factor-dependent output elasticities that, given the choice of $P(E/L) = ac\frac{E/E_0}{L/L_0}$ for the function freely available in Eq. (3.22), satisfy these constraints and the differential Eqs. (3.18), (3.19), and (3.20) are

$$\begin{aligned} \alpha &= a\frac{L/L_0 + E/E_0}{K/K_0}, \gamma = -a\frac{E/E_0}{K/K_0} + ac\frac{E/E_0}{L/L_0}, \\ \beta &= 1 - a\frac{L/L_0}{K/K_0} - ac\frac{E/E_0}{L/L_0}. \end{aligned} \tag{A.46}$$

(The parameter c meant different here than in (3.40)). With these output elasticities, Eqs. (3.8) and (3.9) provide the simplest LinEx function for an early industrialized production system as

$$Y_{fi}(K, L, E; t) = Y_0(t)\frac{L}{L_0}\exp\left[a\left(2 - \frac{L/L_0 + E/E_0}{K/K_0}\right) + ac\left(\frac{E/E_0}{L/L_0} - 1\right)\right]. \tag{A.47}$$

Perhaps this would have become the production function for Germany after 1945 if the Morgenthau Plan to reduce Germany to an agrarian state had been realized. And perhaps this function can also describe the value added of a developing country like Colombia.

A.4.2 Economy Totally Digital

"Economy totally digital" means the production system of a future in which (electrical) energy is available in abundance at all times and a global, close-meshed Internet always works perfectly. This system will ultimately function without human routine work L, if in Industry 4.0 the capital stock sufficiently organises and expands itself, as described by the First Evolutionary Principle of the Factors of Production in Sect. 3.3.1. In this process, the economic weight of capital, α, recedes behind that of energy, γ, to the extent that $\frac{K}{K_0} \rightarrow \infty$ and $\frac{E}{E_0} \rightarrow c\frac{K}{K_0}$ strive. (K_0, L_0, E_0 may well be of the order of magnitude of the factor inputs of highly industrialised countries of the present day).

Whether, under the constraints of the Second Law of Thermodynamics and the space of the biosphere, the "economy totally digital" will ever come about remains to be seen.

Using $\alpha = 1 - \beta - \gamma$, this economy is defined as a system with the asymptotic boundary conditions

$$\beta \rightarrow 0 \text{ and } \gamma \rightarrow 1, \text{ If } \frac{E}{E_0} \rightarrow c\frac{K}{K_0} \rightarrow Z \gg 1. \tag{A.48}$$

The boundary condition for β applies, similarly as before in the highly industrialized countries of the present, to the state of maximum or now total automation. This is described by $E/E_0 \rightarrow cK/K_0$. Furthermore, the boundary condition for γ states that a huge energy input, which produces a huge, self-regenerating and expanding capital stock, remains as the only factor of production and causes growth—similar to the sun causing life on earth.

Simplest factor-dependent output elasticities that satisfy these boundary conditions and the differential Eqs. (3.24), (3.25), and (3.26) when the function freely available in Eq. (3.28) is chosen at $R(E/K) = \frac{1}{c}\frac{E/E_0}{K/K_0}$ are

$$\beta = a\frac{L/L_0}{E/E_0}, \quad \gamma = -a\frac{L/L_0}{E/E_0} + \frac{E/E_0}{cK/K_0}, \quad \alpha = 1 - \frac{E/E_0}{cK/K_0}. \tag{A.49}$$

(The technology parameter a again has a different meaning than before. The technology parameter c is the energy demand of the totally automated, fully utilized capital stock).

With these output elasticities, Eqs. (3.8) and (3.9) yield the simplest LinEx function for the totally digitized future system to be

$$Y_{td}(K, L, E; t) = Y_0(t)\frac{K}{K_0}\exp\left[a\left(\frac{L/L_0}{E/E_0} - 1\right) + \frac{1}{c}\left(\frac{E/E_0}{K/K_0} - 1\right)\right]. \tag{A.50}$$

References

1. Goethe, J. W. von: Vorspiel zur Wiedereröffnung des Theaters in (Bad) Lauchstädt
2. Kümmel, R.: The Second Law Of Economics—Energy, Entropy, And The Origins Of Wealth. Springer, Heidelberg (2011)
3. Fukuyama, F.: The End Of History And The Last Man. Free Press, New York (1992)
4. Hägele, P.: Freche Verse—physikalisch. Limericks über Physik und Physiker, illustriert von Peter Evers, Vieweg, Braunschweig/Wiesbaden (1995) Kapitel Thermodynamik. Der zur Zeichnung gehörende Limerick lautet: "Die Zustandsfunktion Entropie—statistisch seit Boltzmanns Genie. Ich zähl Komplexionen, tu Stirling nicht schonen. Doch richtig versteh' ich sie nie."
5. Napp, V.: Entropie und Entropieproduktion. Erste Staatsprüfung für das Lehramt an Gymnasien—Schriftliche Hausarbeit (theoretische Physik). Bayerische Julius-Maximilians-Universität (1994)
6. Schönwiese, C.-D.: Klimatologie. Ulmer UTB, 3. Aufl., Stuttgart (2008)
7. Wagner, H.-J., Koch, M. K., et al.: CO_2-Emissionen der Stromerzeugung. BWK **59**, 44–52 (2007)
8. Mauch, W.: Kumulierter Energieaufwand—Instrument für nachhaltige Energieversorgung. Forschungsstelle für Energiewirtschaft Schriftenreihe **23**, (1999); kombiniert mit Daten vom Öko-Institut Darmstadt (2006)
9. Stoller, D.: Chinesische Solarzellen haben eine verheerende Umweltbilanz. 2014. http://www.ingenieur.de/Themen/Photovoltaik/Chinesische-Solarzellen-verheerende-Umweltbilanz. Accessed: 15. August 2015
10. Enzyklika *LAUDATO SÍ* von Papst Franziskus über die Sorge für das gemeinsame Haus, hrsg. vom Sekretariat der Deutschen Bischofskonferenz, Bonn (2015)
11. http://www.kulturelleerneuerung.de, Accessed: 25.04.2017
12. Deutsche Physikalische Gesellschaft und Deutsche Meteorologische Gesellschaft: Warnungen vor drohenden weltweiten Klimaänderungen durch den Menschen. Phys. Blätter **43**, 347–349 (1987); s. auch: www.dpg-physik.de/dpg/gliederung/ak/ake/publikationen/dpgaufruf/1987/pdf
13. Deutsche Physikalische Gesellschaft: Energiememorandum. Phys. Blätter **51**, 388 (1995); s. auch: www.dpg-physik.de/veroeffentlichung/stellungnahmen/mem_energie_1995.html
14. Bayerisches Landesamt für Statistik, https://www.statistik.bayern.de/ueberuns/zeitreihen, Accessed: 25.04.2017
15. http://www.skepticalscience.com/iea-co2-emissions-update-2010.html
16. Stern Review Report on the Economics of Climate Change, ISBN number: 0-521-70080-9, Cambridge University Press, 2007; http://www.cambridge.org/9780521700801.
17. Stern, N.: The economics of climate change. Amer. Econ. Rev. **98(2)**, 1–37 (2008)
18. Deutscher Bundestag: Dritter Bericht der Enquete Kommission "Vorsorge zum Schutz der Erdatmosphäre", Drucksache 11/8030. Bonn (1990), S. 855

R. Kümmel et al., *Energy, entropy, creativity*,
https://doi.org/10.1007/978-3-662-65778-2

19. Proops, J.L.R., Faber, M., Wagenhals, G.: Reducing CO_2 Emissions. Springer, Berlin (1993)
20. US National Academy of Sciences: Climate change and the integrity of science. Science **328**, 689–690 (2010)
21. Rahmstorf, S.: Risk of sea-change in the Atlantik. Nature **388**, 825–826 (1997)
22. Rahmstorf, S.: Die unterschätzte Gefahr eines Versiegens des Golfstromsystems. https://scilogs.spektrum.de/klimalounge/die-unterschaetzte-gefahr-eines-versiegens-des-golHrBfstromsystems/HrB. Accessed 03.01.2018
23. Düren, M.: Understanding The Bigger Energy Picture. Springer Briefs in Energy (Open Access), https://doi.org/10.1007/978-3-319-57966-5. Springer Nature, Cham (Schweiz) (2017)
24. von Buttlar, H.: Umweltprobleme. Phys. Blätter **31**, 145–155 (1975)
25. Berg, M., Hartley, B., Richters, O.: A stock-flow consistent input-output model with applications to energy price shocks, interest rates, and heat emissions. New J. Phys. **17**, 015011 (2015) (Open Access). https://doi.org/10.1088/1367-2630/17/1/015011
26. Solow, R. M.: The economics of resources or the resources of economics. Amer. Econ. Rev. **64**, 1–14 (1974)
27. Bundesministerium für Wirtschaft und Energie (BMWi): Global, Innovativ, Fair—Wir machen die Zukunft digital. Öffentlichkeitsarbeit, Berlin (2017)
28. Ebeling W., Freund J., Schweitzer F.: Komplexe Strukturen: Entropie und Information. B. G. Teubner, Stuttgart (1998)
29. Stahl, A.: Entropiebilanzen und Rohstoffverbrauch. Naturwissenschaften **83**, 459 (1995)
30. https://www.drogenbeauftragte.de/presse/pressekontakt-und-mitteilungen/2017/2017-2-quaHrBrtal/ergebnisse-der-blikk-studie-2017-vorgestellt.htmlHrB (Accessed: 10.03.2018)
31. Eichhorn, W., und Solte, D.: Das Kartenhaus Weltfinanzsystem. Fischer Taschenbuch Verlag, Frankfurt/M. (2009); insbes. S. 193
32. H.-G. Hilpert et al.: Aus fremder Quelle. Japans steiniger Weg ins 21. Jahrhundert. Zeitschrift für japanisches Recht **6**, 4. Jg., 139–156 (1998)
33. Samuelson, P. A.: Volkswirtschaftslehre, Band I und Band II. Bund-Verlag, Köln (1975)
34. Franz von Assisi: Sonnengesang von Franz von Assisi. In: Eininger, Ch. (Hrsg.) Die schönsten Gebete der Welt, S. 69. Südwest Verlag, München (1964)
35. Echnaton (Amenophis IV.): Sonnengesang des Echnaton. In: Eininger, Ch. (Hrsg.) Die schönsten Gebete der Welt, S. 94. Südwest Verlag, München (1964)
36. Sieferle, R. P.: Das vorindustrielle Solarenergiesystem. In: Brauch, H. G. (Hrsg.) Energiepolitik, S. 27–46. Springer, Berlin (1997)
37. Heinloth, K.: Energie und Umwelt—Klimaverträgliche Nutzung von Energie. B. G. Teubner Stuttgart, Verlag der Fachvereine Zürich (1993) S. 15; nach: E. Cook, Scientific American **225**, S. 135 (1971)
38. Kümmel, R.: Energie und Kreativität. B.G. Teubner, Stuttgart (1998)
39. Smil, V.: Energy In World History. Westview, Boulder (1994); zitiert nach [36].
40. Institut der deutschen Wirtschaft, Köln
41. Ayres, R. U.: Information, Entropy, And Progress. American Institute of Physics, New York (1994)
42. Ayres, R. U.: Energy, Complexity And Wealth Maximization. Springer International Publishing Switzerland (2016). Dieses Buch vertieft und erweitert—durchaus auch um Alternativen zu aktuellen physikalischen Theorien—die Themen von [41]
43. Gesamtverband Steinkohle: Steinkohle 2014—Herausforderungen und Perspektiven. www.gvst.de/site/steinkohle/Internationale_Energie.htm.
44. The LTI Research Group (Hrsg.): Long-Term Integration Of Renewable Energy Sources Into The European Energy System. Physica-Verlag, Heidelberg (1998)
45. Kondo, J., Inui, T., Wasa, K. (Hrsg.): Proceedings of the Second International Conference on Carbon Dioxide Removal, Kyoto, 1994. Energy Conversion and Management **36**, Numbers 6–9 (1995)

46. Tolba, M. K. (Hrsg.): Our Fragile World—Challenges And Opportunities For Sustainable Development. EOLSS Publishers Co., Oxford UK (2001)
47. www.eolss.net
48. Hohmeyer, O., Ottinger, R. L. (Hrsg.): External Environmental Costs Of Electric Power. Springer, Berlin (1991)
49. Fricke, J., Borst, W. L.: Energie, Oldenbourg, München, Wien, 2. Aufl. (1984)
50. Baehr, H. D.: Thermodynamik, Springer, Berlin, 7. Aufl. (1989)
51. Foulds, L. R.: Optimization Techniques. Springer, Berlin (1981)
52. Blok, K.: Introduction to Energy Analysis. Techne Press, Amsterdam (2006). Dieses Buch vermittelt einen didaktisch gut aufbereiteten Einstieg in die Energieanalyse und verweist auf weiterführende Literatur
53. van Gool, W.: Energie En Exergie. Van Gool ESE Consultancy, Driebergen (1998)
54. Schüssler, U., Kümmel, R.: Schadstoff-Wärmeäquivalente als Umwelbelastungsindikatoren. ENERGIE, **Jahrg. 42**, 40–49 (1990)
55. Kümmel, R., Schuessler, U.: Heat equivalents of noxious substances: a pollution indicator for environmental accounting. Ecol. Econ. **3**, 139–156 (1991)
56. Steinberg, M., Cheng, H. C., Horn, F.: A system study for the removal, recovery and disposal of carbon dioxide from fossil fuel power plants in the US. BNL-35666 Informal Report, Brookhaven National Laboratory, Upton (1984)
57. Fricke, J., Schüssler, U., Kümmel, R.: CO_2-Entsorgung. Phys. Unserer Zeit **20**, 56–81 (1989)
58. Schüssler, U., Kümmel, R.: Carbon dioxide removal from fossil fuel power plants by refrigeration under pressure. In: Jackson, W. D. (Hrsg.), Proc. 24th Intersociety Energy Conversion Engineering Conference, S. 1789–1794. IEEE, New York (1989)
59. Hendricks, C. A., Blok, K., Turkenburg, W. C.: The Revovery of Carbon Dioxide from Power Plants. In: Okken, P. A., Swart, R. J., Zwerver, S. (Hrsg.) Climate and Energy, S. 125–142. Kluwer, Dordrecht (1989)
60. Schuessler, U.: Deponierung und Aufbereitung von CO_2. Phys. Unserer Zeit **21**, 155–158 (1990)
61. https://de.wikipedia.org/wiki/Kraftwerk_Schwarze_Pumpe. Accessed: 9. August 2017
62. Kolm, H.: Mass Driver Up-Date. L5 News **5, No. 9**, 10–12 (1980)
63. O'Neill, G. K.: The Colonization of Space. Phys. Today, **September 1974**, 32–40 (1974)
64. O'Neill, G. K.: The High Frontier—Human Colonies In Space. William Morrow & Co., New York (1977)
65. O'Neill, G. K.: Unsere Zukunft im Raum. Hallwag, Bern (1978)
66. O'Neill, G. K.: The (Low) Profile Road to Space Manufacturing. Astronautics and Aeronautics **16**, Special Section, 18–32 (1978)
67. Steininger, K. W., Lininger, Ch., Meyer, L. H., Muñoz, P., Schinko, Th.: Multiple carbon accounting to support just and effective climate policies. Nature Climate Change **6**, 35–41 (2016), https://doi.org/10.1038/nclimate2867
68. Arbeitsgemeinschaft Energiebilanzen e. V.: Bruttostromerzeugung in Deutschland ab 1990 bis 2016 nach Energieträgern. Stand 11.08.2017
69. Bundesministerium für Wirtschaft (BMWi): Zahlen und Fakten Energiedaten. https://de.wikipedia.org/wiki/Energieverbrauch/cite_note-BMWi_Zahlen-9. Accessed: 12. August 2017
70. Groscurth, H.-M., Kümmel, R.: The cost of energy optimization: A thermoeconomic analysis of national energy systems. Energy **14**, 685–696 (1989). (Diese Arbeit erweitert die Energieoptimierungsstudie *Thermodynamic limits to energy optimization*von Groscurth, H.-M., Kümmel, R., van Gool, W: Energy **14**, 241–258 (1989), um die Kostenoptimierung.)
71. Groscurth, H.-M.: Rationelle Energieverwendung durch Wärmerückgewinnung. Physica-Verlag, Heidelberg (1991); hier wird das Modell*ecco*unter den Bezeichnungen *LEO I* und *LEO II*optimierungstechnisch spezifiziert.
72. Groscurth, H.-M., Kümmel, R.: Thermoeconomics and CO_2-Emissions. Energy **15**, 73–80 (1990) und [71], S. 124–130

73. Kümmel, R., Groscurth, H.-M., Schüßler, U.: Thermoeconomic analysis of technical greenhouse warming mitigation. Int. J. Hydrogen Energy **17**, 293–298 (1992)
74. Bruckner, Th., Groscurth, H.-M., Kümmel, R.: Competition and synergy between energy technologies in municipal energy systems. Energy **22**, 1005–1014 (1997)
75. Bruckner, Th., Kümmel, R., Groscurth, H.-M.: Optimierung emissionsmindernder Technologien. Energiewirtschaftliche Tagesfragen **47**, 139–146 (1997)
76. Ressing, W.: Die CO_2/Energiesteuer—Chance oder Risiko für die Wettbewerbsfähigkeit der deutschen Wirtschaft? Energiewirtschaftliche Tagesfragen **43**, 299–306 (1993)
77. Internationale Energie Agentur (IEA): Weltenergieausblick, Paris (1993)
78. Faross, P.: Die geplante CO_2/Energiesteuer in der Europäischen Gemeinschaft. Energiewirtschaftliche Tagesfragen **43**, 295–298 (1993)
79. Welsch, H., Hoster, F.: General-Equilibrium Analysis of European Carbon/Energy Taxation. Zeitschrift für Wirtschafts- und Sozialwissenschaften **115**, 275–303 (1995)
80. Lindenberger, D., Bruckner, Th., Groscurth, H.-M., Kümmel, R.: Optimization of solar district heating systems: seasonal storage, heat pumps, and cogeneration. Energy **25**, 591–608 (2000)
81. Lindenberger, D., Bruckner, Th., Morrison, R., Groscurth, H.-M., Kümmel, R.: Modernization of local energy systems. Energy **29**, 245–256 (2004)
82. Birnbacher, D.: Intergenerationelle Verantwortung oder: Dürfen wir die Zukunft der Menschheit diskontieren? In: Klawitter, J., Kümmel, R. (Hrsg.) Umweltschutz und Marktwirtschaft, S. 101–115. Königshausen & Neumann, Würzburg (1989)
83. Daly, H.: When Smart People Make Dumb Mistakes. Ecol. Econ. **34**, 1–3 (2000)
84. Rürup, B.: DER CHEFÖKONOM, Newsletter des Handelsblatt Research Institute, 23.02.2018
85. Häring, N.: Mehr Energie! http://research.handelsblatt.com/assets/uploads/AnalyseOekonomischeModelle.pdf, Accessed 26.02.2018
86. https://de.wikipedia.org/wiki/Datei:Crude_oil_prices_since_1861.png; Accessed: 3. Juli 2017
87. Lindenberger, D., Weiser, F., Winkler, T., Kümmel, R.: Economic Growth in the USA and Germany, 1960–2013: The Underestimated Role of Energy. Biophys. Econ. Resour. Qual. (2017) 2:10; https://doi.org/10.1007/s41247-017-0027-y
88. Kümmel, R., Strassl, W.: Changing energy prices, information technology, and industrial growth. In: van Gool, W., Bruggink, J.J.C. (Hrsg.) Energy and Time in the Economic and Physical Sciences, S. 175–194. North-Holland, Amsterdam (1985)
89. Kümmel, R., Strassl, W., Gossner, A., Eichhorn, W.: Technical progress and energy dependent production functions. Z. Nationalökonomie—Journal of Economics **45**, 285–311 (1985).
90. Murray, J., King, D.: Oil's tipping point has passed. Nature **481**, 433–435 (2012)
91. Witt, U.: Beharrung und Wandel—ist wirtschaftliche Evolution theoriefähig?. Erwägen, Wissen, Ethik **15**, 33–45 (2004)
92. Kümmel, R.: Wachstumskrise und Zukunftshoffnung. In: Görres Gesellschaft (Hrsg.) CIVITAS, Jahrbuch für Sozialwissenschaften **16**, S. 11–61, Grünewald, Mainz (1979)
93. Valero, A., Valero, V.: Thantia—The Destiny Of The Earth's Mineral Resources. World Scientifc, Singapore (2015).
94. Bundesministerium für Wirtschaft: Energiedaten '95. Bonn (1996)
95. Meadows, Denis, Meadows, Donella, Zahn, E., Milling, P.: Die Grenzen des Wachstums, dva, Stuttgart (1972)
96. Kümmel, R.: The Impact of Entropy Production and Emission Mitigation on Economic Growth. Entropy **18**, 75 (2016); https://doi.org/10.3390/e18030075 (open access)
97. Hudson, E. H., Jorgenson, D. W.: U.S. energy policy and economic growth, 1975–2000. Bell J. Econ. Manag. Sc. **5**, 461–514 (1974)
98. Berndt, E.R., Jorgenson, D.W.: How energy and its cost enter the productivity equation. IEEE Spectr. **15**, 50–52 (1978)
99. Berndt, E.R., Wood, D.O.: Engineering and econometric interpretations of energy—capital complementarity. Amer. Econ. Rev. **69**, 342–354 (1979)

100. Jorgenson, D.W.: The role of energy in productivity growth. Amer. Econ. Rev. **74/2**, 26–30 (1984)
101. Nordhaus, W.: A Question Of Balance: Weighting The Options On Global Warming Policies. Yale University Press, London (2008)
102. Kümmel, R., Ayres, R.U., Lindenberger, D.: Thermodynamic laws, economic methods and the productive power of energy. J. Non-Equilib. Thermodyn. **35**, 145–179 (2010); https://doi.org/10.1515/JNETDY.2010.009
103. Denison, E.F.: Explanation of declining productivity growth. Survey of Current Business **August 1979, Part II**, 1–24 (1979)
104. Kümmel, R.: The impact of energy on industrial growth. Energy **7**, 189–203 (1982)
105. Tryon, F. G.: An index of consumption of fuels and power. J. Amer. Statistical Assoc. **22**, 271–282 (1927)
106. Binswanger, H.C., Ledergerber, E.: Bremsung des Energiezuwachses als Mittel der Wachstumskontrolle. In: Wolf, J. (Hrsg.) Wirtschaftpolitik in der Umweltkrise. S. 103–125. dva, Stuttgart (1974)
107. Solow, R. M.: Technical Change and the Aggregate Production Function, The Review of Economics and Statistics **39**, 312–320 (1957)
108. Solow, R. M.: Perspectives on growth theory. J. Econ. Perspect. **8**, 45–54 (1994)
109. Robinson, J.: The production function and the theory of capital, Rev. Econ. Stud. **21**, 81–106 (1953–54)
110. Robinson, J.: The measure of capital: the end of the controversy. Econ. J. **81**, 597–602 (1971)
111. Pasinetti, L.: Critique of the neoclassical theory of growth and distribution. Moneta Credito (Banca Nationale del Lavoro Quarterly Review) **210**, 187–232 (2000)
112. Hesse, H., Linde, R.: Gesamtwirtschaftliche Produktionstheorie, Teil I und Teil II. Physica-Verlag, Würzburg-Wien (1976); insbesondere Teil I, S. 11–42 und Teil II, S. 9–30
113. Arrow, K.J., Chenery, H.B., Minhas, B.S., Solow, R.M.: Capital-Labor Substitution and Economic Efficiency. Rev. Econ. Stat. **43**, 225–250 (1961)
114. Uzawa, H.: Production Functions with Constant Elasticity of Substitution. Rev. Econ. Stud. **29**, 291–299 (1962)
115. Lindenberger, D: Wachstumsdynamik energieabhängiger Volkswirtschaften. Metropolis, Marburg (2000)
116. Kümmel, R.: Energie und Wirtschaftswachstum. Konjunkturpolitik **23**, 152–173 (1977)
117. Tintner, G., Deutsch, E., Rieder, R.: A production function for Austria emphasizing energy. In: Altman, F.L., Kýn, O., Wagner, H.-J. (Hrsg.) On the Measurement of Factor Productivities, S. 151–164. Vandenhoek & Ruprecht, Göttingen (1974)
118. Kümmel, R., Henn, J., Lindenberger, D.: Capital, labor, energy and creativity: modeling innovation diffusion. Struct. Change Econ. Dynam. **13**, 415–433 (2002)
119. Winkler, T.: Energy and Economic Growth—Econometric Analysis with Comparisons of Different Production Functions by Means of Updated Time Series for Output and Production Factors from 1960–2013. Master Thesis, Julius-Maximilians-University Würzburg, Faculty of Physics and Astronomy, 2016; als Discussion Paper elektronisch verfügbar.
120. Ayres, R. U., Warr, B.: The Economic Growth Engine. Edward Elgar, Cheltenham UK (2009)
121. Stresing, R., Lindenberger, D., Kümmel, R.: Cointegration of output, capital, labor, and energy. Eur. Phys. J. **B 66**, 279–287 (2008); https://doi.org/10.1140/epjb/e2008-00412-6
122. Institut der deutschen Wirtschaft, Köln, für die 1970er- und 1992er-Daten; The CIA World Fact Book für die 2009er-Daten
123. https://de.wikipedia.org/wiki/Weltenergiebedarf
124. Kümmel, R., Lindenberger, D.: How energy conversion drives economic growth far from the equilibrium of neoclassical economics. New Journal of Physics **16**, 125008 (2014) https://doi.org/10.1088/1367-2630/16/12/125008 (Open access, Special Issue "Networks, Energy and the Economy".)
125. Samuelson P. A., Solow, R. M.: A complete capital model involving heterogeneous capital goods. Quart. J. Econ. **70**, 537–562 (1956)

126. Grahl, J.: private Mitteilung
127. Paech, N.: Regionalwährungen als Bausteine einer Postwachstumsökonomie. Zeitschrift für Sozialökonomie **45/158–159**, 10–19 (2008)
128. Erhard, L.: Wohlstand für alle. Econ, Düsseldorf (1957)
129. Jaeger, W.: Paideia. Die Formen des griechischen Menschen. Walter de Gruyter, Berlin (1933)
130. Haesler, A. J.: Die Doppeldeutigkeit des Fortschritts in der "Philosophie des Geldes". In: Binswanger, H. C., Flotow, P. von (Hrsg.) Geld und Wachstum. Zur Philosophie und Praxis des Geldes, S. 61–78, Weitbrecht, Stuttgart/Wien (1994)
131. Simmel, G.: Philosophie des Geldes. Duncker & Humblot, Leipzig (1900)
132. Schulze, G.: Die Beste aller Welten. Hanser, München/Wien (2003)
133. Gross, P.: Die Multioptionsgesellschaft. Suhrkamp, Frankfurt a. M. (1993)
134. Schmidt-Bleek, F.: Das MIPS-Konzept. Weniger Naturverbrauch—mehr Lebensqualität durch Faktor 10, Knaur, München (2000)
135. Weizsäcker, E. U. von, Hargroves, K., Smith, M.: Faktor Fünf: Die Formel für nachhaltiges Wachstum. Droemer, München (2010)
136. Huber, J.: Nachhaltige Entwicklung. Edition Sigma, Berlin (1995)
137. Braungart, M., McDonough, W.: Einfach intelligent produzieren. Berliner Taschenbuch Verlag, Berlin (2003)
138. Scheer, H.: Solare Weltwirtschaft. Strategie für die ökologische Moderne. Antje Kunstmann Verlag, München (1999)
139. Paech, N.: Nachhaltiges Wirtschaften jenseits von Innovationsorientierung und Wachstum. Metropolis, Marburg (2005)
140. Neirynck, J.: Der göttliche Ingenieur. Die Evolution der Technik. Expert-Verlag, Renningen (2001)
141. Schumacher, E.F.: Small Is Beautiful. Abacus, London (1973)
142. Paech, N.: Postwachstumsökonomik—Wachstumskritische Alternativen zum Marxismus. Aus Politik und Zeitgeschichte (APuZ) **19–20**, 41–46 (2017)
143. Pigou, A. C.: The Economics Of Welfare. Macmillan and Co, London (1920)
144. Kapp, K. W.: The Social Costs Of Private Enterprise. Schocken Books, New York (1950)
145. Binswanger, H.C.: Die Wachstumsspirale. Metropolis, Marburg (2006)
146. Paech, N.: Grünes Wachstum? Vom Fehlschlagen jeglicher Entkopplungsbemühungen: Ein Trauerspiel in mehreren Akten. In: Sauer, T. (Hrsg.): Ökonomie der Nachhaltigkeit—Grundlagen, Indikatoren, Strategien. Metropolis-Verlag, Marburg, 161–181 (2012)
147. Paech, N.: Mythos Energiewende: Der geplatzte Traum vom grünen Wachstum. In: Etscheit, G. (Hrsg.) Geopferte Landschaften. Wie die Energiewende unsere Umwelt zerstört, S. 205–228. Heyne, München (2016)
148. Santarius, T.: Der Rebound-Effekt. Ökonomische, psychische und soziale Herausforderungen der Entkopplung von Energieverbrauch und Wirtschaftswachstum. Metropolis, Marburg (2015)
149. Jevons, W. S.: The Coal Question. An Inquiry Concerning the Progress of the Nation, and the Probable Exhaustion of Our Coal Mines. Macmillan & Co, London & Cambridge (1865)
150. http://www.wbgu.de/fileadmin/user_upload/wbgu.de/templates/dateien/veroeffentlichungen/sondergutachten/sn2009/wbgu_sn2009.pdf.
151. http://uba.co2-rechner.de
152. Paech, N.: Nach dem Wachstumsrausch: Eine zeitökonomische Theorie der Suffizienz. Zeitschrift für Sozialökonomie (ZfSÖ) **47/166–167**, S. 33–40 (2010)
153. Ehrenberg, A.: Das erschöpfte Selbst. Campus, Frankfurt a. M. (2004)
154. Toffler, A.: The Third Wave. Bantam Books, New York (1980)
155. Ostrom, E.: Die Verfassung der Allmende. Jenseits von Staat und Markt. Mohr, Tübingen (1999)
156. Friebe, H., Ramge, T.: Marke Eigenbau. Der Aufstand der Massen gegen die Massenproduktion. Campus, Frankfurt a. M. (2008)

157. http://download.regionalwert-hamburg.de/downloads/Pressemappe_Regionalwert_AG_Hamburg_2016-12-09.zip
158. Kohr, L.: Appropriate Technology. Resurgence **8/6**, 10–13 (1978)
159. Illich, I.: Tools for Conviviality. Harper & Row, Cornell (1973)
160. Mumford, L.: The Myth of the Machine. Secker & Warburg, London (1967)
161. Daly, H.: Steady-State Economics. Island Press, Washington (1977)
162. Reif, F.: Fundamentals Of Statistical And Thermal Physics. McGraw-Hill, New York (1965)
163. Bedford-Strohm, H.: Vortrag am 13. Juni 2012 auf dem 1. Ökumenischen Kirchentag in Höchberg
164. Altmaier P.: Vortrag auf einer Enegiewendekonferenz der Thüringer Wirtschaft in Erfurt, zitiert von der Main-Post Würzburg am 6. November 2012
165. Kümmel, R.: Energiewende, Klimaschutz, Schuldenbremse—Vorbild Deutschland? In: Ostheimer, J., Vogt, M. (Hrsg.) Die Moral der Energiewende, S. 109–133. Kohlhammer, Stuttgart (2014).
166. Merkel, A.: Energie und Rohstoffe für morgen—sicher, bezahlbar, effizient. Wirtschaft in Mainfranken **02/2012**, S. 10–11 (2012)
167. NISA and JNES, 2011: The 2011 off the Pacific coast of Tohoku Pacific Earthquake and the seismic damage to the NPPs. Nuclear and Industrial Safety Agency (NISA), Japan Nuclear Energy Safety Organization (JNES), April 4, 2011, Japan; www.webcitation.org/5xuhLD1j7
168. Gesellschaft für Reaktorsicherheit (GRS): Fukushima Daiichi 11. März 2011—Unfallablauf, Radiologische Folgen, 2. Aufl. GRS, Köln (2013)
169. Heinloth, K.: Die Energiefrage—Bedarf und Potentiale, Nutzen, Risiken und Kosten. Vieweg, Braunschweig/Wiesbaden (1997)
170. Krause, F., Bossel, H., Müller-Reißmann, K.-F.: Energiewende—Wachstum und Wohlstand ohne Erdöl und Uran. S. Fischer, Frankfurt/M (1980)
171. Die Energiewende: unsere Erfolgsgeschichte. Bundesministerium für Wirtschaft und Energie (BMWi), Referat Öffentlichkeitsarbeit (Hrsg.), Berlin (Januar 2017)
172. Energiedepesche 2, 26. Jahrg., Juni 2012, S. 18, und https://tinyurl.com/leitstudie2011.
173. Murphy, D., Hall, C.: Year in review—EROI or energy return on (energy) invested. Ann. N.Y. Acad. Sci. **1185**, 102–118 (2010)
174. Nationale Akademie der Wissenschaften Leopoldina: Bioenergie—Möglichkeiten und Grenzen; Kurzfassung und Empfehlungen. Deutsche Akademie der Naturforscher Leopoldina, Halle/Saale, S. 11f, S. 8 (2010)
175. Umweltbundesamt: Nationale Treibhausgas-Inventare 1990 bis 2015 (Stand 02/2017) und Schätzung für 2016 (Stand 03/2017)
176. Icha, P.: Entwicklung der spezifischen Kohlendioxid-Emissionen des deutschen Strommix in den Jahren 1990–2016. Climate Change **15**, S. 10. Umweltbundesamt, Dessau-Roßlau (2017)
177. Wirth, H.: Aktuelle Fakten zur Photovoltaik in Deutschland (Fassung vom 19.05.2015). Fraunhofer-Institut für Solare Energiesysteme ISE, Freiburg (2015). Dem noch unveröffentlichten englischsprachigen Manuskript dieser Studie wurde die Abbildung 5.1 entnommen.
178. Wirth, H.: Aktuelle Fakten zur Photovoltaik in Deutschland (Fassung vom 21.10.2017). Fraunhofer-Institut für Solare Energiesysteme ISE, Freiburg (2017). Aktuelle Fassung abrufbar unter www.pv-fakten.de. Accessed: 24.10.2017
179. Deutsche Physikalische Gesellschaft: Netzausbau im Rahmen der Energiewende. PHYSIKonkret **18**, 1, Oktober 2013
180. IWR Pressedienst.de, Pressemitteilungen der Energiewirtschaft, Berlin. Accessed: 17.10.2017
181. https://de.wikipedia.org/wiki/Liste_der_größten_Kohlenstoffdioxidemittenten. Accessed: 18.10.201
182. https://de.nucleopedia.org/wiki/Liste_der_geplanten_Kernkraftwerke (Accessed: 24.10.2017)
183. Lindenberger, D.: Volkswirtschaftliche Einordnung des Beitrags der Kohle zur Energietransformation. Energiewirtschaftliche Tagesfragen **67**, 19–22 (2017)
184. Energieagentur NRW, 2017

185. Korrespondenz Wasserwirtschaft **9**, Nr. 5, S. 269 (2016)
186. Kunkel, A., Schwab, H., Bruckner, Th., Kümmel, R.: Kraft-Wärme-Kopplung und innovative Energiespeicherkonzepte. BWK **48**, 54–60 (1996)
187. www.dbi-gti/power-to-gas-methanisierung.html Accessed: 27.10.2017
188. Li, Sh., Gong, J.: Strategies for improving the performance and stability of Ni-based catalysts for reforming reactions. Chem. Soc. Rev. **43**, 7245 (2014)
189. Wang, P., Chang, A. Y., Novosad, V., Chupin, V. V., Schaller, R. D., Rozhkova, E. A.: Cell-Free syntheticbiology chassis for nanocatalytic photon-to-hydrogen conversion. ACS Nano **11**, 6739–6745 (2017). https://doi.org/10.1021/acsnano.7b01142
190. Der Spiegel, 36/2017/S. 81
191. www.br.de/themen/wissen
192. Bundesamt für Migration und Flüchtlinge: Aktuelle Zahlen zu Asyl. Ausgabe September 2017. Siehe auch: Das Bundesamt in Zahlen 2017. (Zugegriffen in der Korrektur: 24.09.2018) Nürnberg. www.bamf.de
193. Bundesamt für Migration und Flüchtlinge: Schlüsselzahlen Asyl 2016. Siehe auch: Das Bundesamt in Zahlen 2017. (Zugegriffen in der Korrektur: 24.09.2018) Nürnberg. www.bamf.de
194. Deutsche Bundesbank: Vermögen und Finanzen privater Haushalte in Deutschland: Ergebnisse der Vermögensbefragung 2014. Monatsbericht, März 2016, S. 67
195. Der Spiegel 25/2017, S. 95
196. Main-Post Würzburg, 26. Juni 2017, S. 5
197. Der Tagesspiegel, 14.01.2018, Politik.
198. Main-Post Würzburg, 15. Dezember 2015, S. 8
199. Main-Post Würzburg, 24. Dezember 2015, "Zitat des Tages"
200. Diamond, J.: Guns, Germs, And Steel. W. W. Norton, New York (1999)
201. Frankfurter Allgemeine Zeitung: Kosten der Migration, aktualisiert am 19.05.2018, https://www.faz.net/aktuell/politik/78-milliarden-euro-fuer-fluechtlingspolitik-bis2022-15598121.html. Zugegriffen (in der Korrektur): 22.05.2018
202. http://www.auswaertiges-amt.de/de/aussenpolitik/laender/kolumbien-node
203. Mapa Ecologica de Colombia, Editorial Educativo KINGKOLOR Ltda, 2002
204. Der Marschall und die Gnade—Roman des Simón Bolivar, Verlag Kurt Desch, München (1954). Das Zitat ist dem Klappentext dieses Buchs entnommen.
205. Nach dem Klappentext zu*Der General in seinem Labyrinth*, Kiepenheuer und Witsch, Köln (1989)
206. http://conflictoarmadointerno2009-1.blogspot.de/2009/05/cronologia-de-las-guerras-en-colombia.html, und die dort angegebenen Quellen.
207. http://www.britannica.com/EBchecked/topic/126016/Colombia/25342/La-Violencia-dictatorship-and-democratic-restoration
208. Caballero, A.: Los Irresponsables. El Tiempo (Bogotá), Lecturas Dominiciales, Pagina 4, Octubre 29 (1972).
209. Echavarría, H.: Miseria y Progreso, Capitulo III: La Revolución Industrial en Colombia, p. 33–40. 3R Editores Ltda, Santafé de Bogotá (1997); s. auch http://www.icpcolombia.org/archivos/biblioteca/49%2D%2D2-Capitulo3
210. Paech, N.: Befreiung vom Überfluss. Oekom, München (2013)
211. Dreier, W., Kümmel, R.: Zukunft durch kontrolliertes Wachstum. Regensberg, Münster (1977), 2. Aufl. 1978.
212. Pfister, Ch. (Hrsg.): Das 1950er Syndrom—Der Weg in die Konsumgesellschaft (Klappentext). Publikation der Akademischen Kommision der Universität Bern. Haupt, Bern (1995)
213. Murphy, D. J., Hall, C.A.S.: Adjusting the economy to the new energy realities of the second half of the age of oil. Ecol. Model. https://doi.org/10.1016/j.ecolmodel.2011.06.022(2011)
214. Trainer, T.: The oil situation: some alarming aspects. thesimplerway.info/OilSituation.htm (Accessed: 24.11.2017)

215. Bundesministerium der Finanzen: perSaldo, Ausgabe 4/1997, S. 6
216. zitiert nach [217]
217. Baron, R.: Competitive issues related to carbon/energy taxation. Annex I Expert Group on the UN FCCC, Working Paper 14. ECON–Energy, Paris (1997). Dieses Dokument erhielt R. K. als einer der beiden Vertreter des Heiligen Stuhls bei den Sitzungen der Subsidiary Bodies of the United Nations Framework Convention on Climate Change vom 20.–31. Oktober 1997 in Bonn.
218. Hannon, B., Herendeen, R.A., Penner, P.: An energy conservation tax: impacts and policy implications. Energy Syst. Policy **5**, 141–166 (1981)
219. Armingeon, K.: Energiepolitik in Europa: Hindernisse umweltpolitischer Reformen. In: [212], S. 377–389
220. Glaser, P.E.: Method and Apparatus for converting solar radiation to electrical power. United States Patent 3, 781, 647, December 23, 1973
221. Glaser, P.E.: Solar power from satellites. Physics Today **February 1977**, 30–38 (1977)
222. Hampf, D., Humbert, L., Dekosrsky, Th., Riede, W.: Kosmische Müllhalde. Physik Journal **17**, 31–36 (2018)
223. Gerst, A.: Gebt eurem Traum eine Chance! Physik Journal **16**, 28–31 (2017)
224. National Space Security Office: Space-Based Solar Power as an Opportunity for Strategic Security. Phase 0 Architecture Feasibility Study, 10 October 2007; www.nss.org/settlements/ssp/library/nsso.htm. Accessed: 15.11.2017
225. www.welt.de/article347436/Atlantis-Flug-beendet-gigantisch-teure-Raumfahrt.html, Accessed: 14.11.2017.
226. Woodcock, G.: Solar power satellite study. L5 News, November 1978, S. 11
227. von Hoerner, S.: Population explosion and interstellar expansion. In: Scheibe, E., Süßmann, G. (Hrsg.) Einheit und Vielheit, Festschrift für Carl Friedrich v. Weizsäcker zum 60. Geburtstag, S. 221–247. Van den Houck & Ruprecht, Göttingen (1973)
228. Boulding, K.: The economics of the coming spaceship Earth. In: Jarret, H. (Hrsg.) Environmental Quality In A Growing Economy, S. 3–14. Resources for the Future, Baltimore MD (1966)
229. Heilbroner, R.: An Inquiry Into The Human Prospect. W. W. Norton, New York (1974)
230. Kammer, H.-W., Schwabe, K.: Thermodynamik irreversibler Prozesse. Physik-Verlag, Weinheim, (1986); S. 60.
231. Kluge, G., Neugebauer, G.: Grundlagen der Thermodynamik. Spektrum Fachverlag, Heidelberg (1993)
232. Georgescu-Roegen, N.: The Entropy Law And The Economic Process. Harvard University Press, Cambridge Mass. (1971)
233. Letters to the Editor. Recycling of Matter. Ecol. Econ. **9**, 191–196 (1994)
234. Karlsson, S: The exergy of incoherent electromagnetic radiation. Phys. Scr. **26**, 329 (1982)
235. van Gool, W.: The value of energy carriers. Energy **12**, 509 (1987)

Printed in the United States
by Baker & Taylor Publisher Services